Megaverse
The Universe of Universes

Why Many Universes? Dynamics. Creation, Features, Expansion, Annihilation.
Escaping via Starships. Life in the Megaverse?

Stephen Blaha Ph. D.
Blaha Research

Pingree-Hill Publishing

ISBN: 978-0-9984575-6-7

Cover Credits

The cover depicts the Megaverse strewn with universes, with starships exiting from our universe,.A self-portrait of Leonardo daVinci furnishes a backdrop for the scene.

Rev. 00/00/01 May 14, 2017

To Margaret

Some Other Books by Stephen Blaha

All the Megaverse! Starships Exploring the Endless Universes of the Cosmos using the Baryonic Force (Blaha Research, Auburn, NH, 2014)

SuperCivilizations: Civilizations as Superorganisms (McMann-Fisher Publishing, Auburn, NH, 2010)

PHYSICS IS LOGIC PAINTED ON THE VOID: Origin of Bare Masses and The Standard Model in Logic, U(4) Origin of the Generations, Normal and Dark Baryonic Forces, Dark Matter, Dark Energy, The Big Bang, Complex General Relativity, A Megaverse of Universe Particles (Blaha Research, Auburn, NH, 2015).

The Origin of Higgs ("God") Particles and the Higgs Mechanism: Physics is Logic III, Beyond Higgs – A Revamped Theory With a Local Arrow of Time, The Theory of Everything Enhanced, Why Inertial Frames are Special, Universes of the Mind (Blaha Research, Auburn, NH, 2015).

New Types of Dark Matter, Big Bang Equipartition, and A New U(4) Symmetry in the Theory of Everything: Equipartition Principle for Fermions, Matter is 83.33% Dark, Penetrating the Veil of the Big Bang, Explicit QFT Quark Confinement and Charmonium, Physics is Logic V (Blaha Research, Auburn, NH, 2015).

The Periodic Table of the 192 Quarks and Leptons in The Theory of Everything: The U(4) Layer Group, Physics is Logic VI (Blaha Research, Auburn, NH, 2015).

New Boson Quantum Field Theory, Dark Matter Dynamics, Dark Matter Fermion Layer Mixing, Genesis of Higgs Particles, New Layer Higgs Masses, Higgs Coupling Constants, Non-Abelian Higgs Gauge Fields, Physics is Logic VII (Blaha Research, Auburn, NH, 2015)

CQMechanics: A Unification of Quantum & Classical Mechanics, Quantum/Semi-Classical Entanglement, Quantum/Classical Path Integrals, Quantum/Classical Chaos (Blaha Research, Auburn, NH, 2016).

All the Universe! Faster Than Light Tachyon Quark Starships & Particle Accelerators with the LHC as a Prototype Starship Drive Scientific Edition (Pingree-Hill Publishing, Auburn, NH, 2011).

From Asynchronous Logic to The Standard Model to Superflight to the Stars; Volume 2: Superluminal CP and CPT, U(4) Complex General Relativity and The Standard Model, Complex Vierbein General Relativity, Kinetic Theory, Thermodynamics (Blaha Research, Auburn, NH, 2012)

New Boson Quantum Field Theory, Dark Matter Dynamics, Dark Matter Fermion Layer Mixing, Genesis of Higgs Particles, New Layer Higgs Masses, Higgs Coupling Constants, Non-Abelian Higgs Gauge Fields, Physics is Logic VII (Blaha Research, Auburn, NH, 2015)

The Origin of Fermions and Bosons,and Their Unification (Pingree-Hill Publishing, Auburn, NH, 2017).

Available on bn.com, Amazon.com, Amazon.co.uk and other international web sites as well as at better bookstores (through Ingram Distributors).

Preface

The universe is a large island of mass and energy. This book gives solid theoretical and experimental reasons to believe that there is more beyond the boundaries of the universe: a Megaverse containing many island universes, of which our universe is one. It then characterizes the features of universes and the Megaverse. In particular it points out that each universe lies in a high dimension Megaverse, which appears to be sixty-four dimensional, based on an extrapolation of features of our universe. Due to the high dimensionality of the Megaverse every point of our universe is surrounded by exterior Megaverse points. The barrier that prevents us from entering the Megaverse is a type of surface tension force – that is infinite for a flat universe such as ours. Thus everything in our universe is confined to it. The book develops the equations for the surface tension force between a universe and the Megaverse. It shows that it is possible for a 'starship' to penetrate the 'veil' of a universe and enter the Megaverse using a large gravitational field (perhaps around a neutron star) to create a curvature 'bubble' and a high starship velocity simulating a high temperature (the Eotvos effect). Together the confining force can be reduced to zero allowing a starship to enter the Megaverse.

The book covers many additional topics: interactions between universes, particle theory in the Megaverse between universes, the Big Bang and the expansion of our universe, a steady state model of the universe based on infalling mass-energy from the Megaverse, the impact of universe surface tension on the expansion of our universe, leaky universes that exchange mass-energy with the Megaverse, the quantum field theory of universes and the Megaverse, the possibility of Megaversian life, and so on.

The WheelerDeWitt equation furnishes us the framework for a quantum Megaverse containing quantum universes. With it we can define a universe particle theory rather like elementary particle theory with interacting universes, and universe-antiuniverse creation from vacuum fluctuations.

The latter part of the book discusses why the Megaverse is likely to be of great interest and why we should go there. We describe starship designs including a design to escape the universe with a slingshot maneuver around a neutron star taking advantage of surface tension force reduction techniques. Aspects of starship communications and life support are also discussed. The author believes that Mankind must eventually go beyond the stars into the other universes of the Megaverse.

CONTENTS

FIGURES and TABLES

1. Our Universe

In this chapter we outline the unified theory of elementary particles and gravitation in our universe. In a sense it provides guidelines that determine the overall form of the Megaverse[1] – the universe of universes. Our universe being one of those universes, the number of which we regard as *very* large, we take to be representative of the set of universes. Its features are similar to those of the other universes. However, in the absence of known constraints on the characteristics of universes we allow the dimensions, and other features, of universes to vary. Thus non-4-dimensional universes may exist. Other universes can have different symmetries, particle masses, and particle interactions. Other universes may also lack a time dimension (making them somewhat dull.) However we will assume that the Megaverse has a time dimension so that universe dynamics becomes possible and interesting.

We discuss the Megaverse, and the features of universes within the Megaverse, in detail in the chapters that follow. In this chapter the focus is strictly on the theory of matter and energy in our universe. Here we outline our theory. Our fundamental theory of the universe is described in detail in *The Origin of Fermions and Bosons, And Their Unification* (referred to as I in this book), which appeared earlier in 2017 as well as earlier books.

1.1 Fundamental Theory of our Universe

Volume I takes the Complex Lorentz group in flat space-time as its starting point and derives the four types (species) of fermions directly: charged lepton, uncharged lepton, up-type quark and down-type quark. It then proceeds to show that the interactions of the Standard Model SU(3)⊗SU(2)⊗U(1) follow directly from Complex Lorentz group geometry. We call this group the Reality group beause it maps complex coordinate systems to real-valued coordinate systems. The existence of Dark Matter leads to an enlargement of the Reality group to SU(3)⊗SU(2)⊗U(1)⊗SU(2)⊗U(1) with the additional SU(2)⊗U(1) factors provide the interactions of Dark Matter. We do not include a Dark Matter strong interaction because 1) if Dark Matter had the known SU(3) Strong Interaction then it wouldn't be 'Dark' and 2) the group R = SU(3)⊗SU(2)⊗U(1)⊗SU(2)⊗U(1) has 16 generators that are all that is needed to transform any flat space-time complex-valued coordinate system to a real-valued coordinate system. Thus Complex Lorentz group space-time geometry requires R and only R—additional factors are superfluous.

Having established the group structure, fermion spectrum and interactions in the one generation case, we then proceeded to introduc fermion generations. We began by noting that there are particle number quanta: Baryon number, Lepton number, and their Dark equivalents.

[1] The Megaverse is a true physical space that is not related to the quantum Megaverse theory of Everritt and others. The possibility that the universes in the Megaverse may each be a Megaverse is not excluded. However, in the author's view the Megaverse concept is a misstep and highly unlikely to be physical reality.

These four number operators lead directly to a U(4) U(4) Generation group that generates four generations of each species of fermion (although only three generations are known at present).

The existence of four generations of each species then leads in turn to another U(4) group—the Layer group—that gives us a Periodic Table of Fermions with three additional replicas of the layer of four generations (with changes in fermion masses) In total we find there are 192 distinct fundamental fermions – 128 quarks and 64 leptons.

Having discovered the riches of the Complex Lorentz group as it leads us from four fermion species to 192 different fermions, we then realize that consistency compels us to accept Complex General Relativity (with the Complex Lorentz group as its flat space-time limit.) It would be illogical if General Relativity were real-valued and flat space-time were complex-valued. So I proceeded to consider Complex General Relativity and discovered that each Complex General Coordinate Transformations can be factored into a real General Coordinate transformation and a complex coordinate transformation. This factorization leads to another U(4) group—that we called the General Relativistic Reality group. This group is distinct from the Complex Lorentz group Reality group R.This new group adds a new set of interactions that affects fermion and boson masses through the Higgs Mechanism. By adding a mass term to each fundamental particle's mass it sets the universe mass scale and yields the principle of equality of inertial and gravitational mass without appeal to Mach or others.

Taken altogether I creates a unified theory of an Extended Standard Model and General Relativity with the symmetry SU(3)⊗SU(2)⊗U(1)⊗SU(2)⊗U(1)⊗U(4)⊗U(4)⊗U(4) plus real-valued General Coordinate transformations. Given this menagerie of interactions I develops a formalism for the 'rotation of interactions' called Ω-symmetry, which introduces a new group symmetry, SU(3)⊗U(64), and a gauge field providing interactions between all 192 fermions. The (broken) symmetry group transformations rotate the eight particle symmetries[2] amongst each other, and, correspondingly rotate the 192 fermions amongst each other. The Weinberg rotation of ElectroWeak theory is a simple example of this form of interaction rotation. Thus the theory is invariant under this symmetry although it is broken to SU(3)⊗SU(3)⊗U(56) as shown in I.

This symmetry completes the unification process. For if one can rotate things into each other, they are in a sense manifestations of the same unified structure.

The unified theory presented in I has a naturalness, and simplicity, that strongly indicate it is correct. From the Complex Lorentz group it extracts the four species of fermions; it grows the four species to four generations based on Baryon and Lepton number conservation (perhaps slightly broken) which yield the U(4) Generation group; the four generations lead to four (broken) conservation laws that in turn yield the four layers of four generations of fermions. Complex General Relativity completes the initial development with a U(4) interactions group.

---

[2] These symmetry groups are $SU(3)_{Color}$, $SU(2)_{Weak}$, $U(1)_{Weak}$, , $SU(2)_{DarkWeak}$, $U(1)_{DarkWeak}$, $U(4)_{Generation}$, $U(4)_{Layer}$, and $U(4)_{RealGrav}$ in the notation of I.

Particle masses are generated by the Higgs Mechanism. Then unifying the interactions we obtain a unified theory with interaction rotation.[3]

Book I concludes by showing that the gravity potential found at solar system distances, at galactic distances, and at inter-galactic distances follow from our theory. It also shows that the theory yields a linear quark potential and the Charmonium potential. Lastly it shows the theory may explain the missing proton spin puzzle, and also the difference in the proton radius of the hydrogen and muonic hydrogen atoms.

1.2 Fermion Periodic Table

The Periodic Table of 192 Fermions was constructed based on a number of reasonable (in the author's view) assumptions.[4] First we required that the fermion species (types) be constructed by complex Lorentz group transformations on a Dirac spinor at rest. This yielded four species of fermions that we identified with charged leptons, neutral leptons, up-type quarks, and down-type quarks. We then found that quarks occupied the $\underline{3}$ of color SU(3). Color SU(3) as well as the normal and Dark Weak interactions of group SU(2), and the normal and Dark electromagnetic interactions were 'derived' from the form of a Lorentz transformation.[5] Thus we found eight species of normal fermions, and four species of Dark fermions (since Dark fermions do not have color interactions – they are SU(3) singlets.) Thus in a generation there are eight normal fermions and four Dark fermions.

Then noting that there are four conserved particle number operators (baryon number,[6] lepton number, Dark baryon number and Dark lepton number) we were led to posit a U(4) Generation symmetry that generated four generations of fermions. Thus we found 32 normal fermions and 16 Dark fermions = 48 fermions in the four generations.

Next we noted that the number of particles in each of the four generations was (almost) conserved. This fact led to another group – the U(4) Layer group – that led to four layers of fermions. The total number of fermions then was found to be 4*48 = 192 fermions (Fig. 1.1).

1.3 Boson Interactions

The eleven interactions[7] that we identified in I and earlier books are

---

[3] Later we will see interaction rotation will play a significant role in our understanding of the Megaverse.

[4] The contents of this, and following, sections, and the figure appear in several of the author's earlier books in 2015 and 2016.

[5] These points all appear in Blaha (2015a).

[6] The author has pointed out in previous books that disparities in measurements of the gravitational constant G could reflect the existence of a baryonic force that, like electromagnetism, leads to a conserved baryon number. Similar comments would apply to other conserved Generation group numbers and almost conserved Layer group numbers.

[7] In book I and earlier books we use a Pseudoquantum quantum field theory formalism that associates a seconf field with each interaction. In the interests of simplicity we use the usual one field formalism here.

1.3.1 The SU(2) Weak Interaction

The Weak interaction SU(2) gauge field is defined as $W^{i\mu}(x)$ for i = 1, 2, 3. Using SU(2) generators we define the matrix form by $W^{\mu}(x) = W^{i\mu}(x)\tau_i$.

1.3.2 The U(1) Electromagnetic Interaction

The U(1) electromagnetic gauge field[8] is defined as $A_E{}^{\mu}(x)$.

1.3.3 The SU(3) Strong Interaction

The Strong SU(3) gauge field is defined as $A_{SU(3)}{}^{i\mu}(x)$ for i = 1, … , 8. Using SU(3) generators we define the matrix form by $A_{SU(3)}{}^{\mu}(x) = A_{SU(3)}{}^{i\mu}(x)T_i$.

1.3.4 The U(4) Generation Group Interaction

The U(4) Generation group[9] generators are denoted G_i and its gauge fields are denoted $U_{\mu i}(X)$. Thus the Generation group terms in covariant derivatives are

$$\mathbf{U^i{}_\mu \cdot G_i}$$

where i = 1, … , 16.

1.3.5 The U(4) Layer Group Interaction

The U(4) Layer group[10] generators are denoted G_{Lk} and its gauge fields are denoted $V_{\mu k}(X)$. Thus the Layer group terms in covariant derivatives are

$$\mathbf{V^i{}_\mu \cdot G_{Li}}$$

where i = 1, 2, …, 16.

1.3.6 The SU(2) Dark Weak Interaction

We assume Dark Weak interactions have the same form as the known SU(2) Weak interactions. the The Dark Weak interaction SU(2) gauge field is defined as $W_D{}^{i\mu}(x)$ for i = 1, 2, 3. Using SU(2) generators we define the matrix form by $W_D{}^{\mu}(x) = W_D{}^{i\mu}(x)\tau_{Di}$ where the generator matrices τ_{Di} are not in the same subspace as the normal SU(2) generators.

1.3.7 The U(1) Dark Electromagnetic Interaction

The U(1) Dark electromagnetic gauge field[11] is defined as $A_{DE}{}^{\mu}(x)$.

---

[8] We separate the ElectroWeak theory into a Weak SU(2) part and an electromagnetic U(1) part. They can be united through an inverse ElectroWeak rotation.
[9] If there are only three generatons of fermions then the Generation group is U(3).
[10] If there are only three generatons of fermions then the Layer group is also U(3).

1.3.8 The U(4) General Relativistic Reality Group Interaction – The Species Group

The U(4) Species group (the General Relativistic Reality Group) interaction gauge field[12] is $A_S{}^{\mu}(x) = A_{R_{flat}}{}^{\mu}(x)$.

This group rotates fermion fields amongst the four normal species and the four Dark species. Each Color subspecies is rotated to a color species of the same color. Intermediate rotations fall into one species or another. See chapter 23 of I for more details.

1.3.9 The 'Interaction Rotation' Interaction - A_Ω

The SU3)⊗U(64) interaction rotation group gauge field[13] is defined as $A_\Omega{}^{ij\mu}(x)$ for $i = 1, \ldots, 8$ and $j = 1, \ldots, 64$. They total 512 gauge field components. Using their 72 generators expressed in the SU(3) 3 representation and the U(64) 64 representation, with matrix denoted $T_{\Omega ij}$, we can define the matrix form as

$$A_\Omega{}^{\mu}(x) = A_\Omega{}^{ij\mu}(x)T_{\Omega ij}$$

where $T_{\Omega ij}$ is a cross product of SU3)⊗U(64) generators. The tensor product generator matrices are 192×192 matrices (since 3*64 = 192). We choose to have a representation with 192×192 matrices due to the 192 fermions in our Periodic Table of Fermions. (See chapter 16 of I.)

This interaction, which we will call the *Ω-interaction*, is described in detail in chapter 31 of I. The gauge fields will correspondingly be called *Ω-fields*.

1.3.10 The Spinor Connection Interaction

The spinor connection used in formulations of vierbein gravity is $B_{\mu ab}(x)$ where a and b are tangent space indices. The vector is combined with γ matrices for use in matrix equations:

$$B^{\mu} = B^{\mu}{}_{ab}\Sigma^{ab}$$

where

$$\Sigma^{ab} = i\,[\gamma^a, \gamma^b]/4$$

Under a local Lorentz transformation S

$$B^{\mu}(x) \to S(x)B^{\mu}(x)S^{-1}(x) - i\,S(x)\partial^{\mu}S^{-1}(x)$$

A simple spin ½ field transforms as

---

[11] We introduce two fields as we did in our article S. Blaha, Phys. Rev. D**10**, 4268 (July, 1974). These fields enable us to define a free electromagnetic lagrangian that is linear in the fields for reasons given elsewhere.
[12] See chapters 22 and 23 of I for a discussion of the origin of this interaction in Complex General Relativity.
[13] The SU(3) factor is *not* color SU(3).

$$(\partial^\mu + i\, B^{1\mu} + i\, B^{2\mu})\psi \rightarrow S(\partial^\mu + i\, B^{1\mu} + i\, B^{2\mu})\psi$$

1.3.11 Real-Valued General Coordinate Connection (Interaction)

The usual gravitational metric field $g_{\mu\nu}$ with

$$\Gamma_{GR}{}^{\lambda}{}_{\mu\nu} = \tfrac{1}{2} g^{\lambda\alpha}(\partial_\mu g_{\alpha\nu} + \partial_\nu g_{\alpha\mu} - \partial_\alpha g_{\mu\nu})$$

We will return to the consideration of particle interactions within the framework of Megaverse universes in later chapters.

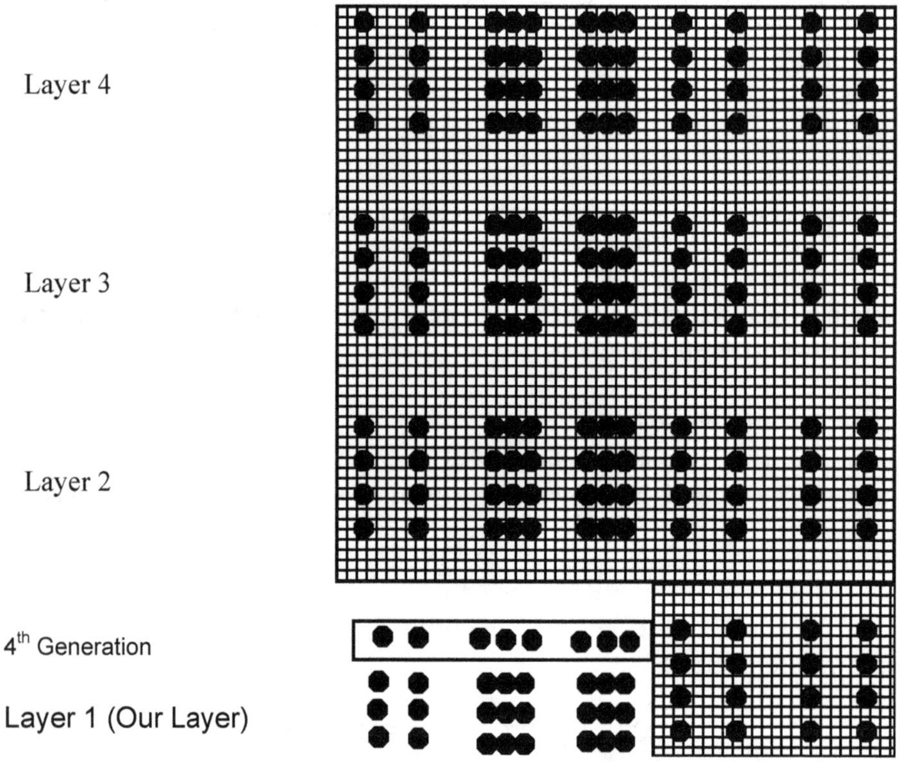

Figure 1.1. Dark parts of the periodic table are 'cross-hatched.' Light parts are the known fermions – with an additional, as yet not found, 4th generation of layer 1 is shown boxed. It is part of 'Dark matter' at present. When found experimentally it will be 'non-Dark.' Normal quarks appear as triplets. Dark quarks are singlets.

2. Theoretical and Experimental Support for the Megaverse and Other Universes

Why are we not content with one universe given its enormous size and variety? It appears that there are important theoretical reasons, and some important experimental observations, that suggest that there is more than one universe.

In this chapter[14] we will discuss theoretical reasons and experimental suggestions of a larger space—the *Megaverse*—that resolves theoretical issues and may address some important astronomical puzzles that have appeared in recent years.

The theoretical issus, which have been subjects of discussion for many years, are:

1. The need for a 'clock' to measure 'time' knowing that it is to some extent relative and local.
2. The need for a 'quantum observer' to complete the understanding of quantum gravity as described by the Wheeler-DeWitt equation and in other efforts to develop a quantum gravity.
3. The need for other universes to provide measuring platforms for quantities beyond the charge and mass of the universe. We think here of the other quantum numbers of particles and particle number operators such as Baryon number.
4. The ultimate source of mass and inertia in our universe.

In Blaha (2015a) and earlier books we have suggested that there are weighty reasons to believe that other universes exist.[15] The existence of other universes is a solution to these problems.

These problems have a source in Quantum Gravity and the interpretation of the Wheeler-DeWitt equation in particular. See appendix 2-A for a discussion of the Wheeler-DeWitt equation and its implications. We now consider the issues raised above.

---

[14] Most of this chapter appears in Blaha (2015a) and in earlier books by the author.

[15] In Blaha (2013a), before the Higgs particle was discovered at CERN we suggested an alternate mechanism was possible if a sister universe existed (making the existence of other universes a reasonable possibility. The Higgs discovery makes the sister universe mechanism unlikely.

2.1 Universe Clocks

Asynchronous Logic provides the equivalent of a clock for the synchronization of processes within large electrical systems such as VLSI chips. Similarly there is a need for a universal clock for our universe. As DeWitt[16] points out in his studies of quantum gravity,

'"The variables … [of the quantized Friedmann model] because of their lack of hermiticity, are not rigorously observable and hence cannot yield a measure of proper time which is valid under all circumstances. … . It is for this reason that we may say that "time" is only a phenomenological concept … If the principle of general covariance is truly valid then the quantum mechanics of everyday usage with its dependence on the Schrödinger equations … is only a phenomenological theory. For the only "time" which a covariant theory can admit is an intrinsic time defined by the contents of the universe itself. Any intrinsically defined time is necessarily non-Hermitean, which is equivalent to saying that there exists no clock, whether geometrical or material, which can yield a measure of time which is operationally valid under *all* circumstances, and hence there exists no operational method for determining the Schrödinger state function with arbitrarily high precision."

The lack of a clock within our universe invalidates quantum mechanics in principle and Quantum Gravity in particular. DeWitt concludes, "Thus [quantum gravity] will say nothing about time unless a clock to measure time is provided."

Unruh[17] also has an issue with the source of time:

"One of the key problems is that of time. We see and experience the world in terms of time. We see things grow, develop, and change. However, time does not enter into the Euclidean formulation of quantum gravity directly. In the usual Hamiltonian formulation, the Hamiltonian for quantum gravity is made up of densities which are the generators, not only of spatial coordinate transformations, but also of temporal coordinate transformations. The content of four of Einstein's equations, namely, the 6 „components, is that these generators are zero. Thus all wave functions are invariant under all spatial and all temporal coordinate transformations. There is nothing in the wave function or the amplitudes which refers to the coordinate t, or the corresponding points of the manifold in any way. How then do we recover the indubitable and ubiquitous experience we have of time? The standard answer is that our experience of time is actually an experience of different correlations between physical quantities in the world. Time is replaced by the readings of clocks. I know that time has changed, not through any direct experience with time, but because the hands of my watch have changed.

Although the implementation of this idea is actually extremely difficult in practice, and although I personally believe that one should formulate one's quantum theory of gravity so as to contain time explicitly, let us nevertheless pursue the consequences of this idea of time as

---

[16] DeWitt, B. S., Phys. Rev. **160**, 1113 (1987).
[17] Unruh, W. G., Phys. Rev. D **40**, 1053 (1989).

defined internally, as the "reading" of a dynamic variable. For an observer inside the theory, his "time" is not the coordinate t. Rather his time is some one of the given dynamic variables of the theory: y or P. Thus although the coupling to the baby universes via the effective action S,. is independent of the coordinates t or x, that does not mean that the observer inside the theory will experience the interactions as being independent of time. For him and/or her, time is one of the dynamic variables and so it can depend on the various dynamic variables of the theory, even if it does not depend on the time coordinate t. In general one would expect the observer to see what looks to him like a time-dependent interaction with the baby universes. At one time, some one of the baby universes may couple strongly to the large universe, while at some other time, another of the baby universes will couple more strongly."

In Blaha (2015a) and earlier books, we suggested the existence of other universes provides a 'clock' in principle for our universe. And being universes, these other universes are an excellent clock. DeWitt points out,

"Because every clock has a "one-sided" energy spectrum, its ultimate accuracy must necessarily be inversely proportional to its rest mass. When the whole universe is cast in the role of a clock, the concept of time can of course be made fantastically accurate (at least in principle) … "

Setting a mass scale using other universes, also sets[18] a time scale and resolves the issue of a clock for our universe. *In principle the existence of other universes validates the role of time in the Copenhagen interpretation of Quantum Mechanics.*

2.2 Quantum Observer

Attempts to create a quantum gravity theory have to confront the need for an *Observer* in any quantum theory within the context of the Copenhagen interpretation. DeWitt points out,

"The Copenhagen view depends on the assumed a priori existence of a classical level to which all questions of observation may ultimately be referred. Here, however, the whole universe is the object of inspection; there is no classical vantage point, and hence the interpretation question must be re-argued from the beginning. While we do not wish to stress this point unduly, since, after all, the Friedmann model ignores the vast complexities of the real universe, it is nevertheless clear that the quantum theory of space-time must ultimately force a deviation from the traditional Copenhagen doctrine."

And Unruh states

[18] For example the Planck time value is set by the Planck mass.

"One of the key features in the interpretation of such transition amplitudes, or wave functions, is the idea that we, as observers are also a part of the Universe as a whole. We, as physical observers, must be describable from within the theory and not as observers external to the theory as in usual quantum mechanics. In usual quantum mechanics, the interpretation is usually given in terms of observers that are outside of the theory. There one makes a split, with the quantum world at one side of the split, and the observer on the other. von Neumann argued that the predictions of quantum mechanics, at least under certain assumptions, are independent of the exact location of that split, but Bohr argued adamantly for the necessity of such a split (classical observers and quantum world). *There is a great difficulty in setting up such a split for physical observers contained within and influenced by a quantum universe,* [itallics added] and for the Universe as a whole, especially including gravity, one cannot argue that the predictions will be independent of where one puts the split. Since all energies interact gravitationally, and our observations are surely energetic phenomenon, the treatment of the energetics of observation as classical would lead to different predictions than if they were treated quantum mechanically. One is therefore forced to devise an interpretation of quantum mechanics in which the observer is part of the quantum system, rather than outside the quantum system.

This means that the interpretation of these transition amplitudes becomes somewhat non-intuitive. One must ask what the system looks like from within, from the viewpoint of an observer who is part of that world, rather than being able to interpret them directly in terms of probabilities for observations made by an external observer."

While the O*bserver* question is addressed by a number of authors, the proposed answers are not entirely convincing. *The existence of other universes provides macroscopic Quantum Observers for our universe.* And our universe provides a macroscopic quantum observer for other universes. Thus the quantum observer issue is resolved.

These considerations lead us to view the existence of other universes as a critical solution to the above problems.

2.3 The Higgs Mechanism is Explainable by Extra Dimensions

The Higgs Mechanism 'explains' (generates) fermion nd boson masses. However the Higgs potential contains a quadratic term with a constant with the dimensions of [mass]. In a sense the Higgs Mechanism trades one mass for another.From where do the Higgs potentials' masses come?

A further explanation is needed is to determine the origin of the "dimensionful" mass terms in the Higgs' particle equations themselves. At present little if any thought has been given to the origin of these terms. We suggested that, excluding a "deus ex machina" source, the only known way to generate these mass terms in the Higgs' equations is through the separation of equations technique of differential equations. This technique requires additional parameters which can only be the coordinates of *extra unknown dimensions*. The best example of the generation of mass terms appears in the Schwarzschild solution of General Relativity where a separation constant, often denoted M, appears that has the dimension of [mass].

Thus extra space-time dimensions would resolve the origin of Higgs potentials' masses. Given extra dimensions it is reasonable to expect that these extra dimensions contain universes. Thus the Megaverse!

2.4 Possible Accretion of Megaverse Matter to Fuel Expansion of Our Universe

If matter is distributed outside of universes in the Megaverse, and if this matter can be accreted to universes by gravitational attraction, then the apparent increasing expansion of our universe may be due to this accretion. In chapter 14 we present a model in which this possibility is realized. If true, then we would have tangible evidence of the residence of our universe in the Megaverse.

2.5 Asynchronous Logic is a Requirement of Universes

By establishing Asynchronous Logic principles[19] as the basis for the existence of universes and for setting the number of dimensions in each universe – four; and basis of fermion particles - iotas – we have found deeper principles of organization for the foundations of physics. The principles built on this foundation serve to enable the coordination of complex physical processes.

Usually we look at particle processes primarily from a space-time perspective: particles collide and produce new particles. We primarily think of the incoming and outgoing particles in a collision. However, considering the set of fundamental particles – and the particle transforming interactions in themselves – neglecting space-time and momentum considerations – leads us to view particles as constituting an alphabet and their interactions as a type of computer grammar.[20] Then the Asynchronicity Principles enable us to bring in space-time in a way that gives us the maximum complexity with the most minimal assumptions. As Leibniz[21] points out our universe has maximal complexity with minimal assumptions.

2.6 The Meaning of Total Quantities of a Universe

The 'external' properties of a universe are normally questioned—for the simple reason that it is assumed that there is no 'outside' of our universe. For example, Misner (1973) asserts:[22]

---

[19] The basis of section 2.3 is described in detail in Blaha (2015a). That book places Physics within a logical framework that is a possible deeper ground for fundamental Physics theory.

[20] This conceptual approach was first described in Blaha (1998) who went on to characterize our universe as one enormous word evolving in time.

[21] See Rescher (1967).

[22] Pp. 457-458.

'There is no such thing as "the energy (or angular momentum, or charge) of a closed universe," according to general relativity, and this for a simple reason. To weigh something one needs a platform on which to stand to do the weighing.'

Misner et al presumes no such platform exists. If there is but one closed universe as most currently believe then one canot measure any totals of a closed universe (which ours may be to be). Yet if we take a more general view that our universe is only one of many then it becomes possible to measure total mass, charge, angular momentum, baryon number, and many other quantities of interest. Indeed, the existence of other universes (within the encompassing Megaverse) opens the door to an understanding of time, mass, energy, and all the other quantities necessary to develop a dynamical theory of universes.

Our new 'rotations of interactions' formalism (described in I) enables us to rotate measurable quantities. These quantities (quantum numbers) furnish a set of totals for our universe such as baryon numbers (normal and Dark), lepton numbers, angular momentum and so on that characterize our universe. We will discuss this in chapter 4.

We will also see that one can then treat universes as 'particles', and develop 'universe dynamics', which might explain knotty problems such as the Big Bang and its precursor (if any). We will do this in subsequent chapters after first considering the possible structure of universes in general in the Megaverse.

2.7 Possible Experimental Evidence for the Megaverse

At first glance it would seem impossible to produce evidence for the existence of other universes. However there are subtle means by which we can 'sense' experimentally 'nearby' universes should they exist. The mechanism would appear to be gravitational effects exerted on objects within our universe by unseen objects of enormous mass. Currently there appears to be three experimental suggestions of the existence of 'nearby' universes and one theoretical argument based on an influx of mass-energy from the Megaverse that may support an understanding of the expansion of our universe.

2.7.1 Great Attractor

One potential support is the discovery of the Great Attractor (at the center of the Laniakea Galaxy Supercluster), and the more massive Shapley Attractor (centered in the Shapley Supercluster)[23]. These attractors contain massive numbers of galaxies and are drawing galaxies over a distance of millions of light years towards them.

If another universe(s) is 'near' our universe it could act as a 'magnet' and draw galaxies towards it to form one or more superclusters which could then act as attractors. Thus attractors might indirectly reveal the presence of other nearby universes—contrary to the expected large

[23] Tully, R. Brent; Courtois, Helene; Hoffman, Yehuda; Pomarède, Daniel, "The Laniakea Supercluster of galaxies". Nature (4 September 2014). 513 (7516): 71–73; arXiv:1409.0880.

scale uniformity of the universe. The only other apparent source of superclusters is chance. Chance seems an unsatisfactory possibility in the present case.

2.7.2 Bright Bumps in Universe Sugesting Collision with Another Universe

A recent study[24] of the residual brightness of parts of the accessible universe found that bright patches appeared if a model of the CMB (Cosmic Microwave Background) with gases, stars and dust was 'subtracted' from the PLANCK map of the entire sky. After the subtraction one would expect only noise spread throughout the sky. However, bright patches were seen in a certain range of frequencies. These anomalies are thought to be a result of our universe colliding with another object – presumably another universe in the Megaverse.

2.7.3 Cold Spot in Universe Suggesting Collision with Another Universe

Another recent study[25] of a huge cold region of the universe spanning billions of light years revealed that this region is not a relatively empty region but rather is similar to in its distribution of galaxies to the rest of the universe. Previous the Cold Spot (an area where cosmic microwave background radiation – the leftover Big Bang radiation is weak – making it significantly colder (0.00015C colder) than the average temperature in the universe 2.73C above absolute zero.)

An analysis of 7,000 galaxy redshifts using new high-resolution data has now shown that the Cold Spot is similar to the rest of the universe. The Durham University group suggested that the Cold Spot might have been caused by a collision between our universe and another Universe. They further suggested that there is only a 1 in 50 chance that it could explained by standard cosmology. could produce this feature

Thus we have another important piece of circumstantial evidence in favor of other universes and thus the Megaverse.

2.7.4 Megaverse Energy-Matter Infusion into Our Universe

In chapter 14 we present a model for an influx of mass-energy from the Megaverse to support the Bondi-Gold-Hoyle-Narlikar Steady State Cosmology, which was originally based on the 'continuous creation of mass-energy' by Hoyle and Narliker. This model explains why the value of Ω makes the universe close to flat. If this model is correct then we would have concrete support for a Megaverse with a low mass-energydensity (sections 14.14 and 14.15) leaking mass-energy into our universe. *More generally, it suggests that universes are surfaces of high mass-energy density in a Megaverse of low mass-energy density – with a ratio of mass-energy densities of the other of* 10^{30}.

---

[24] Ranga-Ram Chary, arXiv.org:/1510.00126 (2015).

[25] T. Shanks et al, Durham University (Australia), Monthly Notices of the Royal Astronomical Society, 2016 .

2.7.5 Conclusion

We conclude that data is beginning to emerge favoring multiple universes and a physical Megaverse in support of the theoretical justifications presented earlier in sectios 2.1 – 2.3..

2.8 Historical Trend Towards Larger Space-Time Structures

Looking back through the history of Mankind's view of the universe we see a clear progression to a larger and larger view. Before the 16th century the earth was the universe. In the 16th century Giordano Bruno (and possibly others) suggested that the stars were suns with many worlds circling them. So our view of the universe expanded to include stars.

Then over time it was noticed that nebulae existed in space. The astronomer, Edwin Hubble, studied the Andromeda nebula with the 'new' Mt. Wilson telescope and in 1929 announced that it was a galaxy composed of stars. Now the universe was conceptually similar to our current view.

Now we seem to have significant theoretical considerations and some suggestive experimental data that lead us to consider the possibility that our universe is not alone—that our universes is but one of many universes in a space we call the Megaverse. This book pulls together much of our earlier work and adds new insights into the nature of the Megaverse. We shall take the Megaverse as fact, extrapolate the form of our universe into the form of other possible universes, and develop a fairly detailed theory of the Megaverse and its resident universes. Then we will consider escaping our universe into the Megaverse and travel to other universes—mindful that such travel will not happen until the distant future. There are many technical bridges to cross before we can travel to other universes.

Given the vastness of our universe one might ask Why travel? We will consider reasons in some detail later. For now, it suffices to say, Because they are there. Mankind has grown and prospered through exploration and exploitation of new territories. Indeed the eminent Historian, Arnold Toynbee, stated that new turf is the source of growth in civilizations. Eventually, in the very distant future, we may need new turf in the Megaverse.

2.9 Other Megaverses?

Does the trend to larger and larger expanses of space suggest that our Megaverse may be but one of many duplicate Megaverses of the same number of dimensions (but diferent orientations)? One cannot decide this question in our present, or likely future, state of knowledge.

However, based on our estimate of the dimension of the Megaverse, and its basis in the geometry and group structure of its interactions, it is reasonable to conjecture duplicate Megaverses would have the same number of dimensions as our Megaverse, and thus have universes with the same number of dimensions as our universe, and thus have a Physics in each other Megaverses' universes similar to the Physics of our universe.

This scenario would appear to be unlikely in the author's view as it appears uneconomical and gives rise to the question How could a plethora of Megaverses arise?

Another scenario, in which Megaverses appear within Megaverses like the toy Chinese nested boxes, also appears unlikely. For it would require a chain of Megaverses of ever increasing dimension, and raise the question of its origin—a question that would never be answerable. Nor would the nested set of Megaverses be experimentally accessible.

Appendix 2-A. Wheeler-Dewitt Equation

This appendix[26] describes the features of the Wheeler-DeWitt equation and some of its implications for quantum gravity in our universe and its extension to the Megaverse.[27]

2-A.1 Quantum Gravity and the Wheeler-DeWitt Equation Extended to Complex Coordinates

In earlier books we developed Quantum Gravity using the weak field expansion and showed that it gave finite results in perturbation theory calculations. There are other attenpts to develop Quantum Gravity without relying on the weak field expansion. These attempts have had some success. Our view is that the existence of one successful approach, in this case the weak field expansion, is proof that Quantum Gravity is viable. Thus we can claim a successful derivation of the unified Extended Standard Model and Quantum Gravity based on space-time geometry.

In this appendix we will consider Quantum Gravity as embodied in the Wheeler-DeWitt[28,29] equation. This equation has many noteworthy points that we will consider below – particularly its extension to complex space-time. It also raises important basic quantum questions: such as "Who is the Observer?" that we addressed earlier. Most of this appendix first appeared in Blaha (2014a).

2-A.2 Analytically Continued Wheeler-DeWitt Equation to Complex Metrics under a Faddeev-Popov Method Restriction

In this section we extend the Wheeler-DeWitt equation for Quantum Gravity to complex coordinates and metrics by analytic continuation (piece-wise if necessary) and impose the condition that metrics must be real-valued using the Faddeev-Popov Method. Our procedure will be to take the Wheeler-DeWitt equation for real-valued metrics (and coordinates), **analytically continue it to the case of complex metrics and coordinates,**[30] and then impose the condition that physically acceptable metrics must be real-valued through use of a Reality

[26] This appendix may be skipped by a reader only interested in the Megaverse.
[27] Extracted from Blaha (2015a) and earlier books.
[28] DeWitt, B. S., Phys. Rev. **160**, 1113 (1987).
[29] Hartle and Hawking also derive the Wheeler-DeWitt equation from a path integral formalism for quantum gravity.
[30] The piecewise analytic continuation of general relativity to complex coordinates and metrics is described in some detail in Blaha (2004). The gist of the continuation is that all equations have the same form after analytic continuation due to a basic theorem of complex variable mathematics that the analytic extension of equations to complex values from real values is unique.

group transformation implemented via the Faddeev-Popov Method. Our motivation is two-fold: we have shown that the form of The Standard Model of Particles can be derived from complex space-time considerations demonstrating that we exist in a "masked" complex space-time; and we have shown the imposition of a restriction on a gauge theory such as gravitation[31] can be implemented using the Faddeev-Popov Method or equivalent.

We start by noting the canonical decomposition of a real-valued metric $g_{\mu\nu}$ is defined by:

$$g_{\mu\nu}(x) = \eta_{\alpha\beta}\, \partial\omega^{\alpha}/\partial x^{\mu}\, \partial\omega^{\beta}/\partial x^{\nu} \tag{2-A.1}$$

with

$$g_{\mu\nu} = g_{\nu\mu}$$

and with inverse

$$g^{\mu\nu} = \eta^{\alpha\beta}\, \partial x^{\mu}/\partial\omega^{\alpha}\, \partial x^{\nu}/\partial\omega^{\beta} \tag{2-A.2}$$

The decomposition of the real-valued metric is

$$g_{\mu\nu} = \begin{bmatrix} -\alpha^2\beta_k\beta^k & \beta_j \\ \beta_i & \gamma_{ij} \end{bmatrix} \tag{2-A.3}$$

$$g^{\mu\nu} = \begin{bmatrix} -\alpha^{-2} & \alpha^{-2}\beta^j \\ \alpha^{-2}\beta^i & \gamma^{ij} - \alpha^2\beta^i\beta^j \end{bmatrix} \tag{2-A.4}$$

where

$$\gamma_{ik}\,\gamma^{kj} = \delta_i^{\,j} \qquad\qquad \beta^i = \gamma^{ij}\,\beta_j \tag{2-A.5}$$

The Wheeler-DeWitt equation for real-valued metrics is

$$(G_{ijkl}\, \delta/\delta\gamma_{ij}\, \delta/\delta\gamma_{kl} + \gamma^{1/2\,(3)}R + 2\lambda\, \gamma^{1/2\,(3)})\Psi({}^{(3)}\mathcal{G}) = 0 \tag{2-A.6}$$

where λ is the cosmological constant, and where the Wheeler-DeWitt metric is

$$G_{ijkl} = \tfrac{1}{2}\,\gamma^{-1/2}(\,\gamma_{ik}\gamma_{jl} + \gamma_{il}\gamma_{jk} - \gamma_{ij}\gamma_{kl}) \tag{2-A.7}$$

The functional derivatives $\delta/\delta\gamma_{ij}$ have several interpretations that are presumably equivalent. DeWitt characterizes them as coordinate independent specifications of the 3-metric. The wave function $\Psi({}^{(3)}\mathcal{G}) = \Psi(\gamma_{ij})$, where ${}^{(3)}\mathcal{G}$ is a geometry, is *not* coordinate dependent. It is invariant

31 Such as real-valuedness.

under coordinate changes. ${}^{(3)}\mathcal{G}$ is a discrete infinity of independent invariants constructed from products of the Riemann tensor and its covariant derivatives.

Hartle and Hawking[32] derive the Wheeler-DeWitt equation from a path integral formalism for quantum gravity. Their path integral can be represented as

$$Z = N \int \delta g(x) \exp(iS_E[g]) \qquad (2\text{-A.8})$$

where S_E is the classical action for gravity and the functional integral is an integral over all 4-geometries. Changing to DeWitt's notation based on the spatial metric γ_{ij} and expressing eq. 2-A.8 in a more explicit form for use in conjunction with the Faddeev-Popov Method we have

$$Z = N \int \sum_{i,j} \prod_x d\gamma_{ij}(x) \exp(iS_E[\gamma]) = N \int D\gamma \exp(iS_E[\gamma]) \qquad (2\text{-A.9})$$

The integrand, being a functional integral over all space, is independent of the coordinates.

The Wheeler-DeWitt equation applies to real-valued metrics $\gamma^{ij}(x)$. We now extend this equation to apply to complex-valued metrics using the local Reality group.[33] In doing this we realize that there are an infinite number of complex-valued metrics in the orbit corresponding to each real-valued metric.

This redundancy can be resolved by realizing that the physical measurement of an invariant interval, and the coordinates from which it is derived, are always real-valued. Yardsticks and clocks can only measure real-valued numbers. Based on this physical principle we can generalize quantum gravity to complex coordinates and metrics by using the Faddeev-Popov method to constrain the set of paths in the quantum gravity path integral eq. 2-A.9. Using the Faddeev-Popov Method the constraint can be expressed in terms of an infinitesimal transformation of the metric to a complex value. Using an infinitesimal Reality group transformation V:

$$[\exp(ia_j(x'',x')U_j)]^\alpha{}_\mu = S(x'',x')^\alpha{}_\mu = \partial x''^\alpha/\partial x'^\mu \qquad (2\text{-A.10})$$

$$V = \exp(ia_k(x'',x')U_k) \cong I + ia_k(x'',x')U_k \qquad (2\text{-A.11})$$

$$V^\dagger = [\exp(ia_k(x'',x')U_k)]^\dagger \cong I - ia_k(x'',x')U_k \qquad (2\text{-A.12})$$

where $a_j(x'',x')$ is the jth real-valued local infinitesimal parameter, and U_k is one of the 16 hermitean generators of the Reality group. (We treat the index on a_k and U_k as lower case and sum on k.) We find the condition fixing the metric to a physical *real* value is:

$$F^{ij}(\gamma(x)) = \text{Im}\, \gamma^{ij}(x) \equiv -\tfrac{1}{2} i(\gamma^{ij}(x) + \gamma^{ij}(x)^*) = 0 \qquad (2\text{-A.13})$$

---

[32] Hartle, J. B. and Hawking, S. W., Phys. Rev. D **28**, 2960 (1983).

[33] The Reality groups are described in I.

where

$$F^{aij}(\gamma^a(x)) = \mathrm{Im}\{(\delta^i{}_m + ia_k(x',x)U_k{}^i{}_m)\,(\delta^j{}_p + ia_k(x',x)U_k{}^j{}_p)\gamma^{mp}(x)\,\} \qquad \text{(2-A.14)}$$

using the infinitesimal form.

The Reality condition eq. 2-A.13 is implemented within the path integral formalism with the Faddeev-Popov Method identity

$$1 = \int D\gamma\,\Delta(F(\gamma))\,\delta(\delta F^{aij}(\gamma^a(x))/\delta a_n(x',x)) = \int D\gamma\,\Delta(F(\gamma))\,\delta(F(\gamma^{ij})) \qquad \text{(2-A.15)}$$

Eq. 2-A.14 yields

$$\begin{aligned}\delta F^{aij}(\gamma^a(x))/\delta a_n(x',x)|_{a=0} &= \mathrm{Im}\{\delta_{kn} iU_k{}^i{}_m\gamma^{mp}(x)\delta^j{}_p + \delta_{kn}\delta^i{}_m i\gamma^{mp}(x)U_k{}^j{}_p]\} \\ &= \mathrm{Re}\{\gamma^{mp}[[U_n{}^i{}_m\delta^j{}_p + \delta^i{}_pU_n{}^j{}_m]\} \\ &= \gamma^{mp}(x)\mathrm{Re}\{U_n{}^i{}_m\delta^j{}_p + \delta^i{}_pU_n{}^j{}_m\} = \gamma^{mp}(x)\xi^{ij}{}_{nmp}\end{aligned} \qquad \text{(2-A.16)}$$

since $\gamma^{mp}(x)$ is made real by the $\delta(\mathrm{Im}\,\gamma^{mp}(x))$ where

$$\xi^{ij}{}_{nmp} = \mathrm{Re}\{U_n{}^i{}_m\delta^j{}_p + \delta^i{}_pU_n{}^j{}_m\} \qquad \text{(2-A.17)}$$

Note $\xi^{ij}{}_{nmp}$ is symmetric in i and j, and becomes effectively symmetric in m and p when combined with $\gamma^{mp}(x)$ in eq. 2-A.16. Calculating $\Delta(F(\gamma))$ we obtain

$$\Delta(F(\gamma)) = [\det \delta F^{aij}(\gamma^a(x))/\delta a_n(x',x)|_{a=0}]^{-1} \qquad \text{(2-A.18)}$$

We can rewrite this Faddeev-Popov determinant as a path integral over an anti-commuting c-number scalar field χ with a ghost Lagrangian:

$$\Delta(F(\gamma)) = \int D\chi^* D\chi\,\exp[i\int d^4x\,\gamma^{1/2}\mathscr{L}_\gamma{}^{ghost}(x)] \qquad \text{(2-A.19)}$$

where

$$\begin{aligned}\mathscr{L}_\gamma{}^{ghost}(x) &= \chi^*(x)[U_{nij}\xi^{ij}{}_{nmp}\gamma^{mp}(x)]\chi(x) \\ &= \chi^*(x)(\mathrm{Re}\,U_{nij})\xi^{ij}{}_{nmp}\gamma^{mp}(x)]\chi(x) &\text{(2-A.20)} \\ &= \chi^*(x)\xi_{mp}\gamma^{mp}(x)\chi(x) &\text{(2-A.21)}\end{aligned}$$

since $\xi_{mp}\gamma^{mp}(x)$ is a c-number:

$$\xi_{mp} = (\mathrm{Re}\,U_{nij})\xi^{ij}{}_{nmp} = 2\,\mathrm{Re}\,U_{npi}\,\mathrm{Re}\,U_n{}^i{}_m \qquad \text{(2-A.22)}$$

using $\mathrm{Re}\,U_n{}^i{}_m = \mathrm{Re}\,U_n{}^m{}_i$ for the U(4) generators in its fundamental representation 4̲.

We then find ξ_{mp} is a diagonal matrix and has the value

$$\xi_{mp} = \xi_m \delta_{mp} \qquad \text{(2-A.22a)}$$

where the diagonal elements are

$$\xi_0 = 8$$
$$\xi_1 = 8$$
$$\xi_2 = 8$$
$$\xi_3 = 8 \qquad \text{(2-A.22b)}$$

and the non-diagonal elements are zero making ξ_{mp} considered as a matrix a multiple of the Identity matrix..

The Faddeev-Popov generated terms when added to the Einstein Action appear to have important ramifications – particularly with respect to the Cosmological Constant. We explore that issue next.

2-A.3 Possible Source of the Cosmological Constant in the Complex Space-time – Faddeev-Popov Constraint Term

The Wheeler-DeWitt equation

$$(G_{ijkl}\ \delta/\delta\gamma_{ij}\ \delta/\delta\gamma_{kl} + \gamma^{\frac{1}{2}\,(3)}R + 2\lambda\,\gamma^{\frac{1}{2}\,(3)})\Psi({}^{(3)}\mathcal{G}) = 0 \qquad \text{(2-A.23)}$$

has the cosmological constant, Λ, as one of its terms. This equation is derived from the Hamiltonian constraint that ultimately follows from the Einstein action. Inserting the Faddeev-Popov term of the previous section in the Einstein lagrangian yields

$$S_E(\gamma) = -(16\pi G)^{-1}\int d^4x\ \gamma^{\frac{1}{2}}\{\ R(x) - 2\lambda + \chi^*(x)\chi(x)\xi_{mp}\gamma^{mp}(x)\} \qquad \text{(2-A.24)}$$

The constant matrix ξ_{mp} is a product of parts of the generators of U(4) given by eq. 2-A.22a:

$$\xi_{mp} = 2\ (\text{Re}\ U_{npi})\ (\text{Re}\ U_{n\ m}^{\ i}) \qquad \text{(2-A.25)}$$

We will now show that the term

$$\Pi = \gamma^{\frac{1}{2}}\chi^*(x)\chi(x)\xi_{mp}\gamma^{mp}(x) \qquad \text{(2-A.26)}$$

upon variation of the metric $\delta\gamma_{\mu\nu}$, gives a term which is approximately a constant cosmological term assuming $\chi(x)$ is approximately constant (with its implied divergence eliminated by renormalization of the path integral), and assuming an almost flat space-time $\gamma_{\mu\nu} \cong \eta_{\mu\nu}$. Varying the metric for Π yields

$$\delta\Pi = \delta\gamma_{\mu\nu}\ \{\tfrac{1}{2}\ \gamma^{\frac{1}{2}}\chi^*(x)\chi(x)\xi_{mp}\gamma^{mp}(x)\gamma^{\mu\nu} - \gamma^{\frac{1}{2}}\chi^*(x)\chi(x)\xi_{mp}\gamma^{m\mu}\gamma^{p\nu}\} \qquad \text{(2-A.27)}$$

The resulting modified Einstein field equation is

$$R^{\mu\nu} - \tfrac{1}{2} g^{\mu\nu}R + \lambda \gamma^{\mu\nu} + \tfrac{1}{2}[\gamma^{mp}(x)\gamma^{\mu\nu} - \gamma^{m\mu}\gamma^{p\nu}]\xi_{mp}\chi^*(x)\chi(x) = -8\pi T^{\mu\nu} \quad (2\text{-A.28})$$

The terms $\tfrac{1}{2}[\gamma^{mp}(x)\gamma^{\mu\nu} - \gamma^{m\mu}\gamma^{p\nu}]\xi_{mp}\chi^*(x)\chi(x)$ appearing above is, or is a contribution to, the cosmological constant assuming a nearly flat universe as our universe seems to be. Thus $\gamma_{\mu\nu} \cong \eta_{\mu\nu}$ to good approximation and $\chi^*(x)\chi(x)$ can be taken to be constant since the time derivative of $\chi(x)$ does not appear in the lagrangian. Consequently the total cosmological constant term is

$$\lambda_{tot}{}^{\mu\nu} \cong \lambda g^{\mu\nu} + 4\,\chi^*(x)\chi(x)g^{\mu\nu} = (\lambda + \lambda_{F\text{-}P})\, g^{\mu\nu} \quad (2\text{-A.29})$$

by eq. 2-A.22b.

Given the somewhat problematic state of our understanding of the cosmological constant it is not impossible that the complexity of space-time leading to $\lambda_{F\text{-}P}$ may be the sole origin of the cosmological constant. In evaluating eq. 2-A.29 we may normalize $\chi^*(x)\chi(x) = 1/4$ by adjusting the overall normalization of the path integral since $\chi(x)$ is time independent. Then if the "bare" cosmological constant is zero we obtain

$$\lambda_{tot} = \lambda_{F\text{-}P} = 1 \quad (2\text{-A.30})$$

by eq. 2-A.29.

If this is true then we have achieved a space-time origin for the cosmological constant rather than an ad hoc origin. The modified Einstein field equation is then

$$R^{\mu\nu} - \tfrac{1}{2} g^{\mu\nu}R + \lambda_{tot}\gamma^{\mu\nu} = -8\pi T^{\mu\nu} \quad (2\text{-A.31})$$

by eqs. 2-A.28 and 2-A.30.

2-A.4 Impact of the Faddeev-Popov Complexity Term on the Wheeler-DeWitt Equation

The Faddeev-Popov term that arises because of the restriction of the metrics and coordinates to real values also impacts on the Wheeler-DeWitt equation since it is derived from the lagrangian via the Hamiltonian it generates. The Wheeler-DeWitt equation changes to

$$(G_{ijkl}\, \delta/\delta\gamma_{ij}\, \delta/\delta\gamma_{kl} + \gamma^{1/2\,(3)} R + 2\lambda\, \gamma^{1/2\,(3)})\Psi({}^{(3)}\mathcal{g}) = 0 \quad (2\text{-A.32})$$

The functional Wheeler-DeWitt equation of eq. 2-A.32 resembles a Klein-Gordon equation.[34] Solutions of this equation can be expressed as path integrals:

[34] Hartle, J. B. and Hawking, S. W., Phys. Rev. D **28**, 2960 (1983).

$$\Psi({}^{(3)}\mathcal{G}, \mathcal{L}_F) = N \int_{\mathcal{C}} \delta g(x) \exp(-I(g, \mathcal{L}_F)) \qquad (2\text{-A.33})$$

where $I(g, \mathcal{L}_F)$ is the effective total Euclidean action for the open universe case. See Hartle and Hawking for a detailed study in the case of a symmetric cosmological constant.

It does not appear that the issues of the Wheeler-DeWitt equation are resolved by the extended Extended Wheeler-DeWitt equation presented here:

- Divergences in integrals in inner products, thus requiring renormalization.
- Negative probabilities in inner products,
- Issues with the requirement of space-like surfaces,
- The frontier divergence singularity.

In chapter 7 we will consider the Wheeler-DeWitt equation for the Megaverse. We will then reconsider the issues of the above formulation.

2-A.4.1 The Universe's Web of Galaxies and the New Faddeev-Popov Terms in the Wheeler-DeWitt Equation

The $12\gamma^{11} + 6\gamma^{22}$ terms in eq. 3.37 introduce a spatial directionality that might be related to the web-like structure of galaxies called the cosmic web that has been found in our universe. The most recent results on this *universal web* which connects all galaxies in the universe are presented by Cantalupo et al in Nature (Jan. 19, 2014). Their results stand in strong contrast to the original expectation of randomly distributed galaxy clusters.

The additional terms in eq. 3.37 strongly suggest broken linear structures that could become web-like as the universe evolved since its beginning.

2-A.5 Is there a Spatial Asymmetry in our Universe Due to the Metric Restriction to Real Values?

The considerations of subsections 3.2 and 3.4 suggest a spatial asymmetry in the universe. Eq. 2-A.33 suggests possible contraction (or slow expansion) in the 1-direction.

$$R^{11} - \tfrac{1}{2} g^{11}R + \lambda_{tot} - 2 = R^{11} - \tfrac{1}{2} g^{11}R - 1 = -8\pi T^{11} \qquad (2\text{-A.34})$$

An analysis[35] of The Sloan Digital Sky Survey data shows the existence of an extremely long, large quasar group powered by ultra-massive black holes that extends 4 billion light-years and is 1.6 billion light-years in breadth in other directions. This is the largest known structure in the universe. It might reflect the asymmetry in the 1-direction with expansion in the 2-direction

[35] Roger Clowes et al, Monthly Notices of the Royal Astronomical Society, Jan. 11, 2013.

and 3-direction but slow expansion in the 1-direction leading to a line of closely spaced quasars in the 1-direction.

Another indication of asymmetry in the universe is a large dense spot in the universe found by the recent European Planck satellite experiment. This region of high density is difficult to understand in conventional theories of our expanding universe.

3. The Extra Dimensions of the Megaverse of Universes

The determination of the dimensions of the Megaverse can only be described as guesswork unless a principle is consistently used to specify the dimensions. In this chapter we will use the known interactions of our universe to determine the Megaverse's dimension.

We will assume that interactions have a dual role in fundamental physics: they determine the dynamics of particles, and they act to determine the dimensions of the Megaverse. The first role is evident within the universe and is therefore obvious from experiment. The second role is external to our universe. It acts to determine the dimensions of the Megavberse.

3.1 Role of Interactions in Determining the Megaverse Dimension D

The motivation for the second role can be discerned from considering a 2-dimensional space, and introducing a simple 1/r potential such as:

$$V = g^2/(x^2 + y^2)^{-\frac{1}{2}}$$

where g is a coupling constant.

One views V as the potential of a force. However the values of V suitably extended to therange $[-\infty, +\infty]$ can be viewed as a third dimension.

With this alternate perspective in mind we take the eight Extended Standard Model vector interactions: E = SU(3)⊗SU(2)⊗U(1)⊗SU(2)⊗U(1)⊗U(4)⊗U(4)⊗U(4) listed in chapter 1 and associate a dimension with the generator index numbers: Thus

$$\begin{array}{lcl} U(1) & \leftrightarrow & 1 \\ SU(2) & \leftrightarrow & 3 \\ SU(3) & \leftrightarrow & 8 \\ U(4) & \leftrightarrow & 16 \end{array}$$

The resulting sum for all eight interactions is the provisional Megaverse dimension[36]

[36] We shall use the symbol D in the remainder of this book for the Megaverse dimension.This choice of D differs from that suggested by I of D = 16. The value of D suggested by I was based on the 16 generators of the Standard Model Reality group. Here we use the 64 generators of the eight vector interactions of E so that we can physically identify the U(64) group part of the Ω-group as the Reality group of the Megaverse. See below.

$$D = 64 \tag{3.1}$$

For consistency, the Megaverse must have complex-valued coordinates since it encompasses the complex-valued coordinates of our universe (although more complicated embeddings can be visualized) We opt for simplicity. In addition the metric of the Megaverse must be Euclidean with a flat-space metric tensor limit consisting entirely of – 1's along the diagonal (using our conventions). For good reason (chapter 14) we view the Megaverse as having at best a very low mass-energy density compared to that of our universe. Thus the Megaverse is essentially flat.

Our calculation of the dimensionality of the Megaverse has the salutary effect of completing the Ω-group defined within our universe. In I we defined this SU(3)⊗U(64).group for the eight vector fields. We will now 'promote' the Ω-group to a symmetry group for Megaverse coordinates (and the spinor indices of fermions).

It also provides the Reality group for the Megaverse since local U(64) can map any (non-degenerate) complex-valued coordinate system to a real-valued coordinate system.

3.2 Megaverse Coordinate Transformations and the *Ω* (Omega) Group of Our Universe

Before extending the Ω-group to Megaverse coordinates, we provide an excerpt from I showing the definition of the group in our universe:

$$\mathbf{A}_I^{1\mu}(x) = (g_1\mathbf{A}_{SU(3)}{}^{1\mu}(x),\ g_2\mathbf{W}^{1\mu}(x)\ ,\ g_3\mathbf{A}_E{}^{1\mu}(x),\ g_4\mathbf{W}_D{}^{1\mu}(x),\ g_5\mathbf{A}_{DE}{}^{1\mu}(x),\ g_6\mathbf{U}^{1\mu}(x),\ g_7\mathbf{V}^{1\mu}(x),\ g_8\mathbf{A}_S{}^{1\mu}(x))$$
$$\mathbf{A}_I{}^{2\mu}_{\ I}(x) = (g_1\mathbf{A}_{SU(3)}{}^{2\mu}(x),\ g_2\mathbf{W}^{2\mu}(x)\ ,\ g_3\mathbf{A}_E{}^{2\mu}(x),\ g_4\mathbf{W}_D{}^{2\mu}(x),\ g_5\mathbf{A}_{DE}{}^{2\mu}(x),\ g_6\mathbf{U}^{2\mu}(x),\ g_7\mathbf{V}^{2\mu}(x),\ g_8\mathbf{A}_S{}^{2\mu}(x))$$

...

Similarly we defined an 8-vector of 64 generators

$$\mathbf{T}_I = (\mathbf{T}_{SU(3)},\ \boldsymbol{\tau}_{SU(2)},\ \mathbf{I}_{U(1)},\ \boldsymbol{\tau}_{DSU(2)},\ \mathbf{I}_{DU(1)},\ \mathbf{G}_{U(4)},\ G_{LU(4)},\ \mathbf{G}_S) \tag{19.17}$$

Then the part of the gauge fields interactions within a covariant derivative corresponding to the eight interactions is

$$\mathbf{A}_I{}^{1\mu}(x)\cdot\mathbf{T}_I + \mathbf{A}_I{}^{2\mu}(x)\cdot\mathbf{T}_I = \mathbf{A}_I{}^{1\mu}{}_k(x)\mathbf{T}_{Ik} + \mathbf{A}_I{}^{2\mu}{}_k(x)\mathbf{T}_{Ik} \tag{31.2}$$

summed over k for each gauge interaction.

...

31.3 Ω-Symmetry, 'Interaction Rotations' for Fermions

We define a column vector ψ containing the 4-spinors of all 192 normal and Dark fundamental fermions in our theory based on four generations in four layers as detailed earlier and in Blaha (2016a), (2016b) and (2016c). Then the Dirac equation has the form:[37]

---

[37] We neglect the real-valued General Relativity interactions.

$$\gamma_\mu D^\mu\psi = \gamma_\mu\{\partial^\mu + i\,[g_\Omega A_\Omega{}^{1\mu}(x) + g_\Omega A_\Omega{}^{2\mu}(x) + \mathbf{A}_I{}^{1\mu}(x)\cdot\mathbf{T}_I + \mathbf{A}_I{}^{2\mu}(x)\cdot\mathbf{T}_I]\}\psi = 0 \quad (31.3)$$

with the Ω-Symmetry gauge field terms $A_\Omega{}^{1\mu}(x)$ and $A_\Omega{}^{2\mu}(x)$ inserted.

In general, the Ω-Symmetry group rotates the 64 = 8+3+2+1+2+1+4+4+4 field components of the eight interactions amongst each other. Thus the Ω-Symmetry group is U(64), to which we add an SU(3) factor to accommodate the 192 fermions in the combined fermion field ψ above, with the result the Ω-Symmetry group is SU(3)⊗U(64). Thus $\mathbf{A}_\Omega{}^{1\mu}(x)$ and $\mathbf{A}_\Omega{}^{2\mu}(x)$ gauge fields each separately have a 192×192 matrix representation of 4-component gauge fields labeled with the index μ. These rotations have 8×64^2 parameters.

The fields in $A_I{}^{1\mu}(x)$, and $A_I{}^{2\mu}(x)$ separately transform under an Ω-transformation consisting of a <u>1</u> representation of SU(3) and a <u>64</u> representation of U(64). They rotate the components of the eight interaction fields $\mathbf{A}_I{}^{1\mu}(x)$ and $\mathbf{A}_I{}^{2\mu}(x)$, each with 64 components, amongst each other. This rotation has 64^2 parameters.

In the case of the fermion field ψ we use the rotations of the <u>3</u> of the SU(3) and the <u>64</u> of U(64). factors to rotate a fermion vector ψ. These rotations have the same 8×64^2 parameters as the $\mathbf{A}_\Omega$ field.

Under an Ω-rotation we find the Dirac equation eq. 31.3 transforms to

$$\gamma_\mu(x)\{\partial^\mu + i\,[g_\Omega A'_\Omega{}^{1\mu}(x) + g_\Omega A'_\Omega{}^{2\mu}(x) + \mathbf{A}'_I{}^{1\mu}(x)\cdot\mathbf{T}'_I + \mathbf{A}'_I{}^{2\mu}(x)\cdot\mathbf{T}'_I]\}\psi = 0 \quad (31.4)$$

where

$$\begin{aligned} A'_\Omega{}^{1\mu}(x) &= C_\Omega(x)A_\Omega{}^{1\mu}(x)C_\Omega{}^{-1}(x) - i\,C_\Omega(x)\partial^\mu C_\Omega{}^{-1}(x)/g_\Omega \\ A'_\Omega{}^{2\mu}(x) &= C_\Omega(x)A_\Omega{}^{2\mu}(x)C_\Omega{}^{-1}(x) \\ \mathbf{A}'_I{}^{1\mu}(x) &= \mathbf{A}_I{}^{1\mu}(x)C^{-1}{}_\Omega(x) \\ \mathbf{A}'_I{}^{2\mu}(x) &= \mathbf{A}_I{}^{2\mu}(x)C^{-1}{}_\Omega(x) \\ \mathbf{T}'_I &= C_\Omega(x)\mathbf{T}_I \end{aligned} \quad (31.5)$$

Note the Species group may cause the γ^μ Dirac matrix to become space-time dependent. (See section 23.1.2 for details.)

The effect of the Ω-transformation is to rotate the gauge fields components. It is accompanied by a rotation of the generator matrices components. Together they define an equivalent formulation of the original Dirac equation and thus the fermion sector. The next section provides an explicit simple example of Ω-transformations – an ElectroWeak theory.

Note that an Ω-transformation causes a change of gauge in the $A_\Omega{}^{1\mu}(x)$ field. The other gauge fields, $\mathbf{A}_I{}^{1\mu}(x)$ and $\mathbf{A}_I{}^{2\mu}(x)$, are 'rotated' but do not undergo a change of gauge. (Each of these other gauge fields do undergo their own particular changes of gauge for their transformation groups.)

The spinor connection field $B^{1\mu}$ (which Weinberg (1972) denotes as Γ^μ on p. 368) is not affected by Ω-transformations since it, as well as the General Coordinate affine connection, is in the gravitation sector, which all fermions (and bosons) experience uniformly.

We now add Megaverse coordinate and spinor transformations to the Ω-group.[38] Generalizing eq. 31.4 for the transformation of the total set of 192 fermions. We define

$$\psi'(y') = C_\Omega\psi(y)$$

where C_Ω is now a local U(D) transformation that transforms coordinates and spinors. Then the transformed massless Dirac-like equation has the form

[38] In this discussion we assume there are D – 1 spatial coordinates and one time coordinate y^D.

$$\gamma_\mu(y')\{\partial'^\mu + i\,[g_\Omega A'_\Omega{}^{1\mu}(y') + g_\Omega A'_\Omega{}^{2\mu}(y') + \mathbf{A'}_I{}^{1\mu}(y')\cdot\mathbf{T'}_I + \mathbf{A'}_I{}^{2\mu}(y')\cdot\mathbf{T'}_I]\}\psi'(y') = 0 \qquad (3.2)$$

The gauge fields transform similarly to eq. 31.5 above in y coordinates. Thus we have a Megaverse Extended Standard Model generalized to Megaverse coordinates. Note that the μ coordinate index now extends from 1, … , D.

3.3 Fermion Spins in the Megaverse for D = 64

The spinors in free field expansions of $\psi(y)$ have $2^{D/2}$ column entries.[39] The D 'Dirac' γ-matrices have $2^{D/2} \times 2^{D/2}$ rows and columns. The lowest fermion spin is

$$s_M = (2^{D/2-1} - 1)/2 \qquad (3.3)$$

For D = 64 these quantities are:

Number of spinor components:
$N_{MRC} = 2^{D/2} = 4{,}294{,}967{,}296$ (3.4)

Spin:
$s_M = 2{,}147{,}483{,}647/2 \quad \approx 1{,}073{,}741{,}824 \approx 1$ billion

Spin Range
$s_{Mz} = -2{,}147{,}483{,}647/2, -2{,}147{,}483{,}647/2 + 1, \; \ldots \;, +2{,}147{,}483{,}647/2$

At first glance the range of spins gives one pause for thought. However, universes are not small generally. Consequently a great range of spins is understandable.

[39] See Weinberg (1995) p. 216.

4. General Properties of the Megaverse

If one wishes to have a depiction of the Megaverse it would seem likely that the universes within it would be scattered in a fashion similar to galaxies within our universe although on a much larger distance scale and in multiple dimensions. Fig. 4.1 portrays a 2-dimensional depiction of the Megaverse. In this chapter we overview properties of the Megaverse and its universes. Much of this material previously appeared in Blaha (2015a) and in earlier Physics and starship travel books.

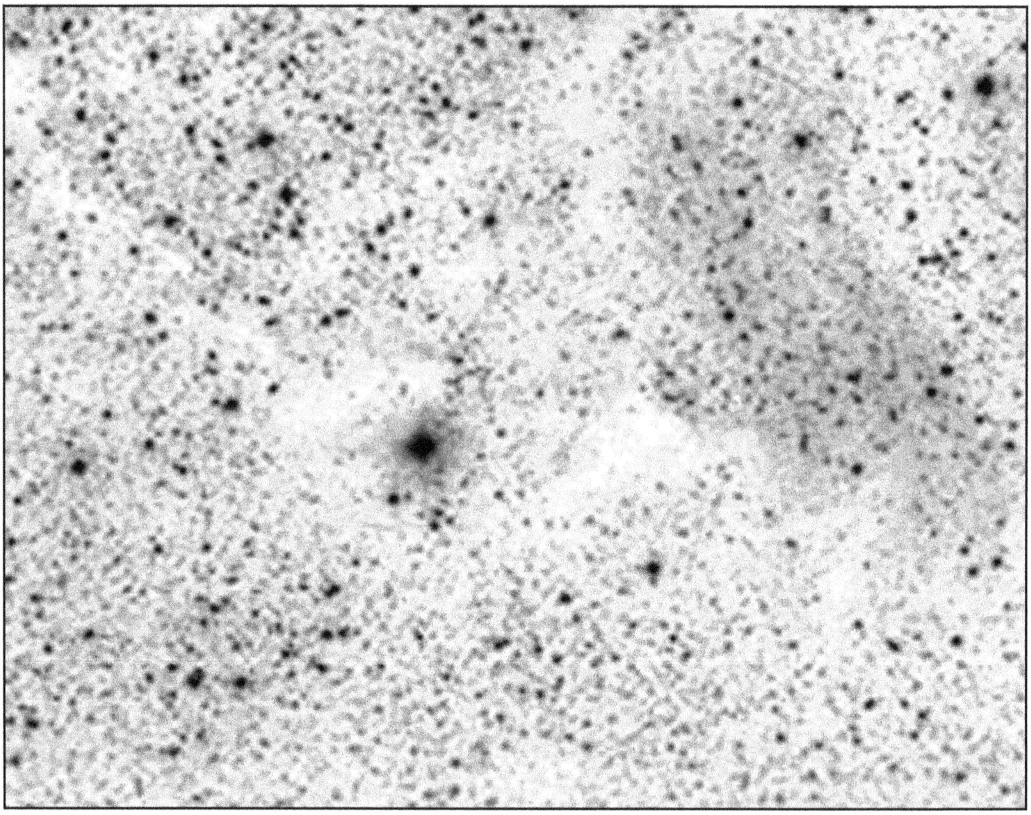

Figure 4.1. Illustrative depiction of the Megaverse, representing universes as black spots.

4.1 Likely Features of the Megaverse

There are a number of features of the Megaverse that we believe to be true:

4.1.1 Megaverse Dimension

The Megaverse has a dimension that can be expressed as D complex/dimensions. Chapter 3 bases the dimensionality of the Megaverse on the interactions of our universe.[40]

4.1.2 Megaverse Curvature

We will assume the Megaverse has a form of gravitation that gives it curvature. The sources of Megaverse gravitational curvature are the masses and motions of the universes within it. Megaverse gravitation will be discussed later in detail. Since the Megaverse density[41] is quite likely to be much lower than the density of our universe, the Megaverse is probably not closed.

4.1.3 Megaverse Time Dimension

We will assume that the Megaverse has one complex time dimension denoted y^D for the simple reason that the absence of a time dimension would make make the Megaverse static.

4.1.4 Megaverse Forces

In addition to Megaverse gravitational effects between universes, the universes within the Megaverse have long range forces such as the baryonic force that cause them to interact with each other.

4.1.5 Megaverse Dynamics

Physical constants and particle masses may have different values in Megaverse space outside of universes. The baryonic, Dark baryonic, and other long range forces cause universes to exhibit motions, and interactions, on a relatively long time scale.

4.1.6 Megaverse Vacuum Fluctuations

Megaverse Vacuum fluctuations may be a source of the generation of universes. This type of event might account for the Big Bang. The time scale for the persistence of universes generated by a vacuum fluctuation is likely to be an extrapolation of vacuum fluctuation persistence within our universe.

---

[40] We will assume that all the universes in the Megaverse are 4-dimensional and the same set of interactions within them. Six-dimensional and higher dimension universes may be present in the Megaverse.

[41] Chapter 14.

4.1.7 Megaverse Matter and Chemistry

The existence of many more dimensions in the Megaverse suggest that multi-dimensional forms of matter and energy could exist between universes. As a result Megaverse atoms, compounds and Chemistry will be very different and much more varied than that of our universe. If such matter exists in the Megaverse then 'mining' such matter for use in our universe—would give us exotic new compounds and Chemistry that would be partially inside, and partially outside, of our universe.

This possibility makes venturing into the Megaverse economically and scientifically desirable since such materials could not be created within our universe.

4.2 Features of Universes Within the Megaverse

We know of our universe from the 'inside.' However the features of our universe from a Megaverse perspective are not at all certain. Im this section we will describe the Megaverse view of a universe's properties.

We shall assume a universe is a closed or open surface within the Megaverse of much higher mass-energy density than the Megaverse's mass-energy density by perhaps as much as a factor of 10^{30}. Chapter 14 describes a model of Megaverse mass-energy inflow into our universe exemplifying this feature.

4.2.1 Universe Area and Mass

The mass of a universe is an important property since mass is one of the sources of Megaverse gravitation, and interaction between universes.

While a universe is not believed to be a black hole (although Hawking has recently jokingly? suggested that our universe may be a black hole, and even more recently suggested black holes are not quite black holes – grey?), there are general qualitative similarities that lead us to consider the possibility that the four laws of black holes[42] may apply in part (or their entirety) to universes. In particular the 2nd law states

$$dM = \kappa dA/8\pi + \Omega dJ \tag{4.1}$$

where dM is the change in "mass/energy," A is the area of the Black Hole (universe), Ω is its angular velocity and J is the angular momentum.[43] From eq. 4.1 it appears we can reasonably define a "mass" for a universe in terms of a universe's area:

$$M = \kappa A/8\pi \tag{4.2}$$

[42] Wald, R. M., "The Thermodynamics of Black Holes", *Living Reviews in Relativity* **4** (6): 12119 (2001).

[43] Although the angular momentum of a universe is not measurable if there is only one universe (as DeWitt argued in a quote in chapter 2), the existence of multiple universes within the Megaverse enables the relative angular momentum of a universe to be determined.

This definition seems to capture the physics of universes that could be used in developing a dynamics of universes as we do later. It allows us to escape the dilemma of having zero total energy for universes that would preclude treating universes as particles in the Megaverse and then developing a Megaverse dynamics of universe-particles. Later we will also define a mass for a universe that is time dependent.

4.2.2 Confinement of universes due to 'Surface Tension'

We assume that the other universes have the same physics as our universe with the possible difference that they may have differing coupling constants and particle masses. As we will discuss later every point of a universe in a higher dimensional Megaverse has Megaverse points in any neighborhood of the point (except for neighborhoods strictly within the universe). Thus we confront the question: what keeps mass-energy at points in a universe or is there leakage from the universe into the Megaverse?

If there is no leakage into the Megaverse then, since each point in the universe is part of a Megaverse surface, one can only assume that there is a barrier to movement into the Megaverse. Taking a note from fuid dynamics and viewing the Megaverse as one 'material' and the universe as a different 'material' we can view the barrier as due to 'surface tension.'[44] The Megaverse must exert a force confining the contents of the universe to within it.

The surface tension[45] of a universe γ satisfies the relation

$$\gamma = W/\Delta A \tag{4.3}$$

where γ is expresased in erg/cm^2, W is the Work, and ΔA is the Area upon which the work is exerted. The pressure exerted by the surface tension for a 'spherical' surface area is

$$\Delta p = 2\gamma/R \tag{4.4}$$

where R equals the radius of curvature of the surface. Eq. 4.4 embodies the concept that the surface tension force equals the pressure difference at the surface.

If the universe is flat then the surface pressure approaches ∞ giving confinement of fields and particles to the universe.

$$R \to 0 \quad \text{implies } \Delta p \to \infty \tag{4.5}$$

Thus we have the theorem:

---

[44] See Landau (1987).

[45] A useful analogy: the Megaverse is a pool of water; a universe is a denser oil bubble within it. Surface tension caused by the cohesiveness of the oil molecules in the bubble makes it spherical (confines it to a spherical shape). Similarly a universe (denser than the Megaverse) is 'confined' within the Megaverse.

Theorem: A universe has no leakage of fields or particles if it is exactly flat.

This theorem is particularly interesting in the case of our universe. It appears to be flat (or very close to flat). The flatness of our universe may be the reason no leakage of fields or particles from our universe has been detected to high accuracy.

If a universe is found with a non-zero radius of curvature then one can expect that some fields and particles may emerge from it into the Megaverse.

While a zero radius of curvature prevents the exit of fields and articles from a universe, it does not prevent the entry of mass-energy into the universe from the Megaverse. Thus our Continuous creation model of chapter 14 may be relevant. Entry is possible; exit is forbidden in this case.

In section 16.3 we consider the possibility that starships can exit into the Megaverse, due to the curvature of our universe in the vicinity of a small massive neutron star (or similar small, very massive object with a large gravity). Basically the curvature induced by the body lowers the surface tension force to a finite value. High temperature can also lower surface tension as we discuss in section 16.3.

4.2.3 Fields Emanating from a Universe

If the curvature of the universe is zero (open universe) then no fields emanate from it. If the curvature of the universe is non-zero (closed universe) then fields may 'leak' into the Megaverse. Then continuity conditions between a universe field and its Megaverse equivalent becomes of interest.

4.2.4 Megaverse Dynamics due to Universe Fields

The universes within a Megaverse have dynamical motions in the Megaverse due to the forces exerted between them as well as Megaverse gravity. We will address dynamical issues in chapter 10.

4.2.5 Types of Universes

It appears that the motions of a universe lead to it being classified as one of four universe species: normal universe traveling at a sublight speed, a tachyonic universe traveling at a speed greater than the speed of light, a universe with a complex-valued velocity whose magnitude is below light speed, and a universe with a complex-valued velocity whose magnitude is above light speed.[46] The possibility of tachyonic universes can be explicitly seen in Appendix 2-A in the solutions of the Wheeler-DeWitt equation.

---

[46] While the four types of universe particle have real-valued energy, their 'energy' need not be real-valued since universes, unlike fundamental fermions, can 'decay.'

Based on the implications of the Wheeler-DeWitt equation it appears reasonable to treat universes as second quantized particles with either spin[47] s_M (fermions) or spin zero – *universe particles* – with angular momentum, as well, due to their motions through the Megaverse. Higher spins are not ruled out.

In discussing universes we can further separate them into closed universes and flat universes. Our universe is close to flat but may be curved. Other universes may have more extreme curvature that make them particle-like in the normal sense of the word.

Universe particles are described in chapter 10.

4.2.6 Universe Coordinate Systems vs. Megaverse Coordinate Systems

Each universe occupies a region of the Megaverse that is an open set in Megaverse coordinates and a closed set in the coordinates of the universe. Within the boundaries of a universe one can use either Megaverse coordinates or the curved coordinates of the universe.

See chapter 7.6.3 (and appendix A) for a detailed discussion. The transformation between Megaverse and universe coordinates for a quantum field necessitates Bogoliubov transformations to preserve particle interpretations of states.

4.2.7 Impact of Universe Rotation of Interactions on the Megaverse

In Blaha (2017b) and earlier books we advocated a group of transformations of eight interactions. This transformation group will also apply in the Megaverse. Chapter 3 gives a detailed discussion of the Ω-Group of interaction rotations.

4.2.8 Megaverse Gravitation and Free Matter

We assume that Einstein's theory of Gravity (or something similar to it) applies when generalized to the many-dimensioned Megaverse. Chapter 8 discusses Megaverse gravitation.

Just as our universe has matter and radiation between galaxies, it seems reasonable to assume that 'free' matter and radiation could exist in the Megaverse outside of universes. Such mass-energy would have two roles: to gravitationally affect the dynamics of the Megaverse and the motion of universes within it, and to possibly fuel the expansion of universes. The expansion of our universe may be due to an influx of matter and energy from the external Megaverse. Many years ago Hoyle and Narlikar considered the possibility of 'continuous creation of matter.' We suggest that an influx of Megaverse matter may be the actual source. We consider this possibility in chapter 14, which contains a paper (unpublished) by this author approximately seven years ago

Thus we arrive at a view of the Megaverse of universes that is analogous to our universe of galaxies.

---

[47] See chapter 3 for the definition of s_M.

4.2.9 Expansion of Universes

Our universe expanded from the Big Bang to its current size and is still expanding. It is likely that other universes are or have undergone similar expansions. According to section 4.2.2 there is likely an infinite surface tension, forsce barrier at the boundary of our universe. How can the universe have expanded, and continue expanding, under such conditions. We see a two phase expansion of the universe.

For a period of time after the Big Bang the universe did not have an infinite surface tension preventing expansion into the Megaverse due to an effect discovered by Eőtvos – a temperature dependence of the surface tension force. Eőtvos pointed out a critical temperature T_c existed that caused the surface tension force to decline as the temperature increased:

$$\gamma V^{2/3} = k(T_c - T) \tag{4.6}$$

where k is the Eőtvos constant, and V is the volume of the universe (the liquid 'drop') Assuming a spherical universe, the volume is

$$V = 4\pi R^3/3. \tag{4.7}$$

Thus

$$\gamma = (4\pi/3)^{-2/3} k(T_c - T)R^{-2} \tag{4.8}$$

For very high temperatures such as existed after the Big Bang $T > T_c$ and thus γ would be negative indicating that there was an outward pressure from the universe into the Megaverse promoting expansion. Thus in the high temperature period after the Big Bang the surface tension force favors expansion of the universe.

After this phase, the surface tension γ is positive. The Megaverse is then superficially 'impeding' expansion. However, the surface tension pressure of the Megaverse causes leakage *into* the universe causing its mass to increase, and by eq. 4.2 causes its radius and volume to increase – Expansion! – due to the accretion of Megaverse mass-energy.

The above scenario is supported by the two phase model suggested by section 14.15 consisting of a Big Bang expansion model (chapters 11 – 13), and a mass-energy accretion model (chapter 14). Note as the radius of curvature goes to zero, eq. 4.8 suggests an increasing surface tension pressure 'pushing' particles into the universe.

Thus a complete universe expansion scenario is evident based on surface tension physics.

4.2.10 Universe Generation from Vacuum Fluctuations

Vacuum fluctuations could generate universe-antiuniverse pairs. Antiuniverses would have "negative" quantum numbers. We view universes/antiuniverses, when created (Big Bangs), as 'ultra-small' particles of mass-energy that can proceed to expand to great size. If they are generated by a vacuum fluctuation they will, after possibly a certain time, recombine into the vacuum.

The creation of universes as vacuum fluctuations can account for the extreme abundance of matter in our universe – a long discussed issue. See chapter 10 for details.

4.2.11 Universes as Black Holes

It is conceivable that a universe could be so dense and confined that it would be effectively a Black Hole. In this case the internal quantum numbers of the Black Hole universe would be inaccessible and only its mass, velocity and angular momentum would be observables.

4.2.12 Life in Other Universes

It is likely that life exists in some if not all of the universes of the Megaverse. If the physical constants, laws, and masses of the interior of a universe are similar if not identical to ours then the possibility of intelligent species, even human-like species, is very likely. This possibility is an important motivation for humanity to reach for the Megaverse as we discuss later. See chapter 27 on Megaverse life.

5. Origin: A Pristine Megaverse

5.1 Nature of an 'Empty' Megaverse

We begin with a characterization of the Megaverse as a D-dimensional space that is initially without universes but has a uniform, low density of mass-energy and the eleven interactions that we described in I. Due to interactions, and fluctuations, universes are created, of which our universe is one.

We assume that the universes that are created have matter, energy, and interactions that we have found in our universe. We also assume the interactions in our universe are present in the Megaverse and other universes. The Megaverse may have more interactions: a U(D) Reality group which partly appears in our universe as the Ω-group, a D-dimensional Lorentz group, and a D-dimensional General Relativity.

So there is a plainness[48] in all universes. A notable new feature of the Megaverse would be pairs of universes created by vacuum fluctuations since the Megaverse as stated earlier is quantum. One universe of the pair might have a preponderance of matter while the other universe might have a preponderance of anti-matter.

Thus the 'empty' Megaverse may be the 'father/mother' of all universes with their contents originally coming the Megaverse.

5.2 The Origin of Universes

Despite the above, the origin of the universes of the Megaverse is a difficult question.[49] The simplest general answer postulates all universes are generated ultimately from vacuum fluctuations of the baryonic field. Subsequent to their creation they may combine or fission into new generations of universes IF the time for the recombination of universes into the vacuum is sufficiently long. This scenario resolves the problem of the original creation of universes within an initially empty Megaverse. In this section we will examine creation from vacuum fluctuations, universe collisions, the coalescence of universes, and the fission of universes. We return to the issue in more detail in chapter 10.

---

[48] The plainness would be mitigated by the appearance of life, especially intelligent life, in universes. The variety of possible forms of life bestows uniqueness on universes. See Feinberg (1980) for a view of the variety of possibilities for life in just about any environment. Experiments *on earth life* are now showing the validity of their thesis.

[49] Most of the material in this section and following sections of this chapter appears in Blaha (2015a) and earlier books.

5.2.1 Creation of Universes through Gauge Field Fluctuations

One of the most exciting questions in Cosmology is the origin of our universe. The conventional view is that it originated in a Big Bang from an infinitesimal point in space. The source of the Big Bang and the prior state of the Cosmos, if there was one, is the subject of much speculation. Based on the particle interpretation of the Wheeler-DeWitt equation it is reasonable to consider the possibility that our universe originated in a vacuum fluctuation.

Our formulation of universe particle theory[50] provides for the generation of universe particle – anti-particle pairs as a vacuum fluctuation. We can view a universe particle as having a substantial excess of baryons, N, as we see in our universe. Its anti-universe at the time of creation (the Big Bang point) is its "mirror image" having the same number of anti-baryons (baryon number: –N) so that baryon number is conserved by the fluctuation event.

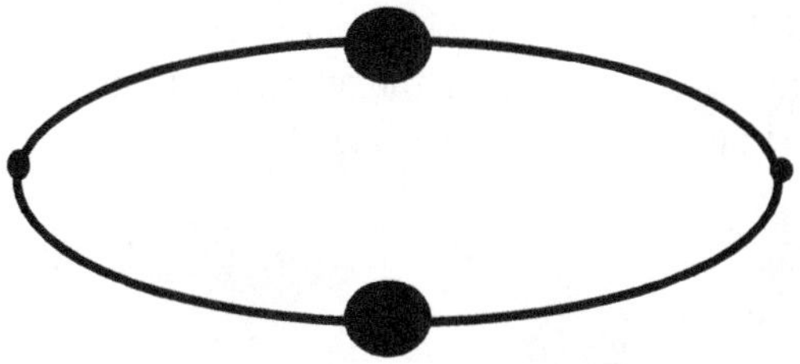

Figure 5.1. Generation of a universe – anti-universe pair as a vacuum fluctuation.

A small coupling constant value could lead to an extremely long lifetime for universes generated by a fluctuation. Thus the 13.7 billion year life of our universe is not unreasonable. Its lifetime can be extremely long. The probability of the creation of universes by vacuum fluctuations would be correspondingly small.

5.2.2 When Universes Collide: Coalescence of Universes

Universes moving in the Megaverse may collide through chance, or due to the forces. When universes collide[51] several possibilities present themselves:

1. They can graze each other distorting each other's shape and internal baryon distribution through the baryonic force while maintain their individual identity.

2. They can intermix with both the baryonic, electromagnetic, and gravitational forces causing a redistribution of their masses. They may separate afterwards or may coalesce into a single universe. One result of this may be lopsided universes. *Our universe*

[50] Chapter 10. Also see Blaha (2015a) and earlier books by the author.
[51] The experimental results of section 2.7 may be relevant.

appears to be lopsided. Some cosmologists believe this is due to a near collision of our universe with another shortly after the Big Bang.

5.2.3 Fission of Universes

Under certain circumstances the distribution of matter in the universe may lead to the fission of a universe into two separate universes. Our model lagrangian (chapter 10) supports this possibility for universe particles. The detailed mechanism of the fission process is not specified by the model.

5.2.4 Fission of "Normal" universes

The fission of normal (non-tachyon) universe particles in our universe particle model is depicted in the Feynman diagram in Fig. 5.2. The sum of the masses of the output universe particles is usually less than the original universe particle mass. However if the fission takes a long time and the masses are time dependent then the combined masses of the produced universe particles may exceed the original universe's mass.

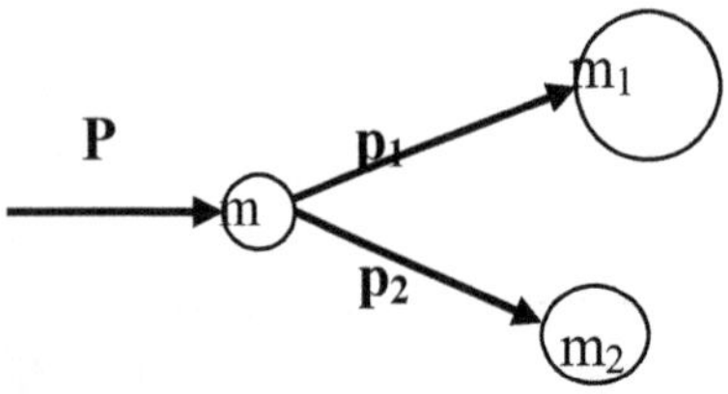

Figure 5.2. Fission of a universe particle into two universe particles.

5.2.5 Tachyon Universe Particle Fission to More Massive Universe Particles

In Blaha (2007a) we showed that a tachyonic (faster than light) particle could fission into particles of larger mass. In this section we will show that a tachyonic universe particle may fission into more massive universe particles. This phenomenon is of particular interest because it enables tachyonic universes to spawn in a new novel way not previously considered in discussions of the origin of universes.

When a particle or a universe particle fissions (decays) one normally expects that the masses of the particles or universe particles produced by the decay to be smaller than the mass of the original particle or nucleus. In the case of tachyonic (faster-than-light) elementary particles or universe particles a much different possibility is present: a tachyon can decay into heavier tachyons. We will consider the specific case of a tachyon universe particle decaying into two universe particles whose total mass is greater than the original. (See Fig. 5.3.)

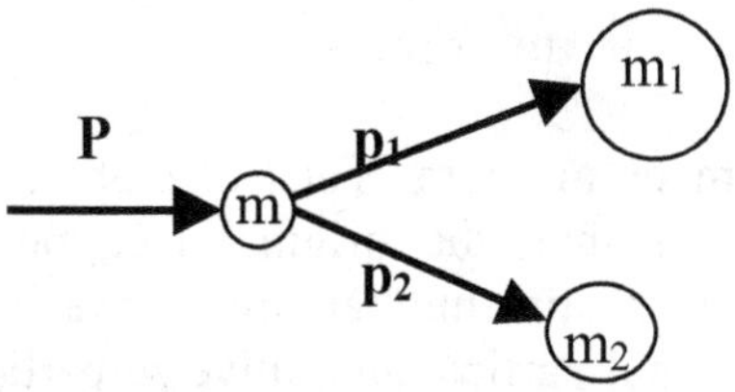

Figure 5.3. Two universe particle decay of a tachyon universe particle.

We will assume the initial tachyon universe particle has zero energy ($p^0 = 0$) and thus the tachyons universe particles emerging from the decay also have total universe particle energy zero. The analysis is based on conservation of total universe energy and momentum in the Megaverse. The below discussion applies to D-dimensional space with (D – 1)-dimensional spatial coordinates.

Momentum conservation implies

$$\mathbf{P} = \mathbf{p_1} + \mathbf{p_2} \tag{5.1}$$

Since all energies are zero

$$(cP)^2 = (c\mathbf{P})^2 = m^2 \tag{5.2}$$

$$\begin{aligned}(cp_1)^2 = (c\mathbf{p_1})^2 = m_1^2 \\ (cp_2)^2 = (c\mathbf{p_2})^2 = m_2^2\end{aligned} \tag{5.3}$$

where $P = |\mathbf{P}|$, $p_1 = |\mathbf{p_1}|$, and $p_2 = |\mathbf{p_2}|$. If we now square eq. 5.1 and then use eqs. 5.3 we obtain

$$m^2 = m_1^2 + m_2^2 + 2m_1m_2 \cos\theta \tag{5.4}$$

where θ is the angle between the emerging universe particles momenta $\mathbf{p_1}$ and $\mathbf{p_2}$. Eq. 5.4 has a number of interesting cases:

Case $\theta = 0$:

$$m = m_1 + m_2$$

The masses of the outgoing universe particles sum to the mass of the original tachyon universe particle.

Case $\theta = \pi/2$:

$$m^2 = m_1^2 + m_2^2$$

The masses of each outgoing universe particle tachyon is less than the mass of the original tachyon universe particle.

Case $\theta = \pi$:

$$m^2 = (m_1 - m_2)^2$$

In this case either $m_1 > m$ or $m_2 > m$. Thus one of the outgoing tachyon universe particles has a greater mass than the original tachyon universe particle. Mass is effectively created from the spatial momentum of the initial universe particle. This process is the inverse of normal particle and universe particle fission where the sum of the outgoing masses is always less than the original particle's mass and the difference is mass converted into energy in the form of additional photons.

This last case, where one of the outgoing universe particles is more massive than the original universe particle, is not just for $\theta = \pi$. Since

$$\cos\theta = (m^2 - m_1^{\ 2} - m_2^{\ 2})/(2m_1m_2)$$

we see that the sum of the outgoing universe particle masses is always greater than the original tachyon universe particle *mass (except when* $\theta = 0$*)* since

$$\cos\theta = 1 + [m^2 - (m_1 + m_2)^2]/(2m_1m_2) \leq 1$$

and thus

$$[m^2 - (m_1 + m_2)^2]/(2m_1m_2) \leq 0$$

Note $m = m_1 + m_2$ only if $\theta = 0$.

Since we can transform the above discussion to the case of universe particle tachyons having non-zero Megaverse energy using an ordinary D-dimensional Lorentz transformation the discussion in this subsection is general.

We therefore conclude that when a tachyon universe particle decays into two tachyon universe particles the sum of the masses of the produced tachyon universe particles is greater than the mass of the original tachyon universe particle except if the angle between the momenta of the produced tachyon universe particles is zero. In that case the sum of the masses of the produced tachyon equals the mass of the original tachyon universe particle and the produced universe particles overlap.

5.3 Multiple Megaverses?

We have assumed that there is one Megaverse. However if more dimensions exist then there is no reason why more Megaverses would not exist. The possibility of a countable infinity of Megaverses is not excluded.

Other Megaverses would not be accessible unless some interaction existed that straddled the Megaverses. Such an interaction would have to be super weak to avoid a significant interplay between the Megaverses – perhaps even a mixing of Megaverses.

A possible inter-Megaverse force would have to be a function of all the coordinates of all Megaverses. Based on an analogy with the multi-universe baryonic force the form of the inter-Megaverse force could be an abelian or non-abelian gauge field. The inter-Megaverse force charge would be perhaps the number of universes in a Megaverse minus the number of anti-universes, N, or a Megaverse mass such as the universe mass of eq. 4.2. This quantum number may or may not be conserved. The inter-Megaverse force may or may not have an associated conservation law similar to the conservation law for electric charge.

5.4 Embeddings of 'Child Universes'

5.4.1 Embedding 4-Dimensional Curved Space-time in the Megaverse

Although The Standard Model is defined in a flat 4-dimensional space-time it appears that the universe may be curved and closed. We can embed our universe within a Megaverse. If our universe were a 4-dimensional real universe, then its metric $g_{\mu\nu}$ has 10 independent components that are determined by the Einstein dynamic equations. Thus we could embed a real universe as a surface in a real 10-dimensional flat space as Eddington (1952) pointed out.

However a 4-dimensional *complex* universe has a metric $g_{\mu\nu}$ with 16 complex-valued components. If one embeds our complex 4-dimensional curved space-time within a Euclidean flat space, then the flat space must be at least a 20-dimensional real space specified by twenty equations:

$$z_i = f_i(x) \tag{5.5}$$

where x is a complex 4-vector in our universe[52] (as in chapters 2 and 3 of Blaha (2012b)) and z_i for i = 1, …, 20 are the coordinates of a flat space point. The functions f_i map our universe into a 20 dimensional real, flat space as a complex 4-dimensional surface (an 8-dimensional real-valued surface.)

We now point out that D-dimensional Megaverse space should have be a D-dimensional *complex* space for consistency with the complex-valued space-time of our universe.[53] In Blaha (2012b) with invariant interval we defined an invariant distance with

$$ds^2 = g_{ij}dz^i dz^j \tag{5.6}$$

where the Megaverse metric satisfied

$$g_{ij} = g_{ji} \tag{5.7}$$

[52] Our universe would then be a complex 4-surface within a 16-dimensional complex flat space.
[53] The present book uses a different dimension Megaverse with D dimensions having complex-valued coordinates as discussed in chapter 3.

The metric tensor $g_{\mu\nu}$ of the universe can be defined in terms of Megaverse coordinates with

$$g_{\mu\nu} = \partial f_j/\partial x^{\mu}\, \partial f_j/\partial x^{\nu} \tag{5.8}$$

with an implied sum over the subscript j.[54] From these definitions we can develop Megaverse General Relativity.(See chapter 6.)

5.5 Ultra-high LHC Proton-Proton Collisions Resemble Heavy Nuclei Collisions

Our earlier discussion of tachyonic universes interactingor decaying to produce heavier universes is buttressed by new particle interaction data from CERN. New 7 TeV proton-proton collision experiments by the CERN ALICE Collaboration[55] have revealed that the end products of these collisions resemble the end products of heavy nuclei collisions.

Tachyonic particles can transform momentum into mass as pointed out in section 5.2.5. Since d and s[56] quarks are tachyonic (See I for the details) and are present in proton-proton collisions, it is likely that the tachyonic mass production mechanism, at least partly, causes pp collisions to simulate heavy nuclei collisions. The quasi-free nature of the quark-gluon plasma generated in ultra-high energy pp collisions lends weight to this interpretation.

The ALICE results may indicate tachyonic mass creation from the momentum of d and s quarks. Normal u and c quarks do not have a tachyonic nature and thus cannot generate mass from momentum. Their collisions generate mass from energy.

[54] This form represents a return to the complex form of invariant intervals of Blaha (2004). An alternate form of complex invariant intervals and metrics presented in Blaha (2012b) does not appear to be of interest after some study by this author.

[55] J. Adam et al, Nature Physics (2017). DOI: 10.1038/nphys4111. Data from LHC run 1.

[56] Enhance production of s quarks is another feature of ultra-high energy pp collisions according to the ALICE collaborate
on.

6. Universes Within the Megaverse

In this chapter we describe the embedding of universes within the Megaverse. Much of this chapter appears in several earlier books by the author such as Blaha (2015a).

As stated earlier, we define a universe to be a closed or open surface in the Megaverse with a much higher mass-energy density than the Megaverse.

There are two types of boundaries for a universe embedded in a space of larger dimensions. First there is a boundary of the universe determined by treating the universe as a surface in the space. Secondly, there is another type of universe boundary defined by the observation that any neighborhood – not strictly within the universe – of every point of the universe has an infinite number of points of the enclosing Megaverse space.[57] Or, every point has a neighborhood with Megaverse points within it. Thus *each point of a universe is on a boundary of the universe due to the larger dimensions of the Megaverse space* within which it resides. Fig. 6.1 schematically illustrates these neighborhoods for any universe point for a universe contained within a higher dimensional Megaverse.

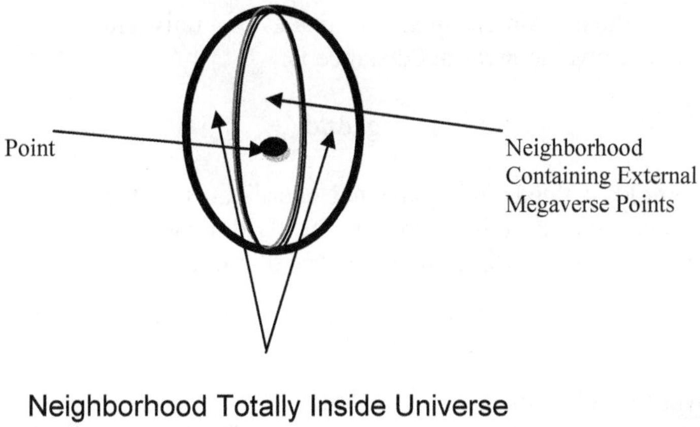

Figure 6.1. Schematic diagram of a 3-dimensional projection of 'orthogonal' neighborhoods of a point within a universe with one neighborhood strictly within the universe and the other neighborhood containing both universe and external Megaverse points in general. The 'orthogonal' circles around the point differentiate between the two types of neighborhoods.

---

[57] Any neighborhood of any point in the universe – with all its points strictly within the universe – has an infinite number of points within the universe. We assume the neighborhood is so small that the curvature of the universe's space can be neglected.

We will consider the classical physical nature of each type of boundary in this chapter and the quantum field nature in chapter 7.

6.1 Embedding Universes Within the Megaverse

In chapter 4 we developed a view of the general nature of universes in the Megaverse. We now examine the general nature of the Megaverse itself. Consistency *suggests that complex-valued coordinates universes, such as our universe, require an embedding in a complex-valued Megaverse* space. In chapter 3 we determined the dimension of the Megaverse, D, from the dimensions implied by interactions within our universe and from a need to extend 'Ω-group symmetry to Megaverse coordinates.

We now assume the Megaverse has an analytic, complex, symmetric metric g_{ik} which satisfies

$$g_{ij} = g_{ji} \tag{6.1}$$

for i, j = 1, 2, … , D.

The embedding equations for our curved complex-valued universe {x} within the complex, D-dimensional Megaverse {y} are:

$$z_i = f_i(x) \tag{6.2}$$

where x^μ are the complex-valued 4-dimensional coordinates of a universe.

The infinitesimal Megaverse invariant distance is

$$ds^2 = g_{ik}dz^i dz^k \tag{6.3}$$

where the z_i are the complex-valued D-dimensional coordinates of the Megaverse. g_{ik} is the complex-valued Megaverse metric tensor, ds is the D-dimensional invariant Megaverse distance,[58] and the complex-valued, 4-dimensional metric of a universe has the form

$$g_{\mu\nu} = \partial f_j/\partial x^\mu \, \partial f_j/\partial x^\nu \tag{6.4}$$

with an implied sum over the subscript j.

The picture that we paint of the Megaverse is that of a complex, D-dimensional space containing (perhaps) countless universes, some of which may be (almost) flat like ours, and some of which may be curved, open or closed 4-surfaces. *We assume universes are closed, or curved and open, or flat.*

The Megaverse does have gravitation due to universe masses, particle masses, the Two Tier[59] field Y^μ, the Baryonic Field B^μ, the Dark baryonic field $B_D{}^\mu$ and the other known gauge

[58] We note that ds^2 is complex and this might be found troubling. A Reality group transformation would yield the absolute value of ds^2 and thus $ds_{physical} = ||ds^2|^{½}$ would be the physical real-valued invariant distance. This remark also applies to 4-dimensional complex General Relativity.

fields (from I). *We will assume all universes are complex, 4-dimensional although universes with a larger number of dimensions are possible.*

6.2 A Universe as a Surface in a Higher Dimensional Space

Every universe is assumed to be a closed or open surface in the Megaverse. Thus any motion within a universe is confined to the universe unless compelled to exit the universe by collisions with Megaverse particles or by external Megaverse forces. All fields defined within a universe in universe coordinates are confined to the universe due to surface tension as described earlier.

If the surface of a universe is as pictured above, then all interactions within the universe are effectively confined to within the universe. *However, since a confined universe has a charge, spin, mass, Baryon Number and Dark Baryon Number, Megaverse interactions exist between universes sourced in these quantities.Also universes with low surface tension may 'leak' fields and particles into the Megaverse.*

6.3 The 'Point' Boundary at Every Point of a Universe within the Higher Dimensional Megaverse

Every point in our universe is "infinitely" close to points of the Megaverse.

A universe occupies a region within the Megaverse. However because it is a lower dimension surface within the Megaverse the neighborhood of every point within a universe has an infinite number of Megaverse points that are not within the universe.

One might think that particles within a universe could then 'slip' into the Megaverse outside the universe with ease. However that is not the case. The law of momentum conservation compels particles and interactions within a universe to be confined to the universe. More importantly, the Megaverse surface tension of a flat universe confines particles and fields within a universe.[60]

The only possible ways that a particle could exit from a universe are 1) if the particle collides with a particle with a momentum, some of whose components are in Megaverse dimensions extraneous to the universe's dimensions, or 2) a particle within the universe experiences forces with components in Megaverse dimensions extraneous to the universe. We shall consider the second possibility later when we consider a mechanism for a starship to exit our universe (chapter 16). The first possibility exists if the Megaverse has a very low matter density outside of universes (chapter 14). The fact that this phenomena has not been observed implies the Megaverse matter density is extremely low.

---

[59] Described in I.

[60] A useful analogy: the Megaverse is a pool of water; a universe is a denser, oil bubble within it. Surface tension caused by the cohesiveness of the oil molecules in the bubble makes it spherical (confines it to a spherical shape).

Thus conservation of momentum for particles and interactions, and surface tesion, effectively confines particles within a universe even though the neighborhood of every point of a particle's trajectory contains an infinity of Megaverse points exterior to the universe. Similarly every interaction within a universe is confined when expanded in a fourier series (assuming free fields) in universe coordinates.

It is also possible for a Megaverse particle to enter[61] *a universe through perhaps a collision that results in the particle being within the universe with momentum also solely within the universe. Thus the 'point boundary' of universes is porous. Particles can enter/exit a universe under appropriate conditions.*

We conclude the 'point boundary' of a universe is not a barrier although surface tension force controls the entry/exit of particles and fields. We consider an exit mechanism for a starship in chapter 16.

6.4 External Properties of Universes Within the Megaverse

Some major characteristics of our universe are its baryon number, its surface area, and its spin. In the case of our universe the preponderance of matter over anti-matter yields an enormous baryon number. In chapter 15 of Blaha (2015a) and chapter 16 of this book we consider the possibility of a baryonic gauge field that would produce a force between universes, between a universe and baryons in another universe, and between baryons within a universe. This field, if it exists, and there is some evidence for it, would account for *baryon number conservation* and also would have significant consequences on the galactic level within our universe. Naturally it must be significantly weaker than the force of gravity. But the plenitude of baryons in our universe gives it importance on large scales.

There is also the possibility that a baryonic force could induce movement, collisions, universe amalgamations, and other effects between universes in the Megaverse making universes a peculiar form of particle on a Megaverse scale. We consider these possibilities in chapter 10.

6.5 Objects Straddling a Universe-Megaverse Boundary

When a starship, or some other extended object, is entering/exiting a universe at some velocity the question of the state of the object arises It is partially in and partially out of the universe. We know that the object being 4-dimensional will continue to be 4-dimensional, barring effects of forces that might "twist" parts of the object into additional dimensions.

There is also the more subtle quantum effects on the object due to the possibility of different quantizations of the particles in a universe and the Megaverse. Quantizations in different coordinate systems might result in different physical interpretations of matter. (For discussions of this possibility see the author's paper in Appendix A and the references cited

[61] In chapter 14 we consider a model that supplies a mechanism for the Hoyle-Narlikar continuous creation theory for the expansion of our universe. This model 'creates' mass-energy as an inflow from the Megaverse.

therein.) In section 7.6.3 below we show that one can quantize using a form of Pseudoquantization that preserves (unitarily equivalent) particle interpretations in a universe and the Megaverse.

Thus extended objects can be partly in a universe, and partly in the external Megaverse.

7. Quantum Field Theory of the Megaverse and Universes

In this chapter we describe the Megaverse as a quantum entity within the framework of the Wheeler-DeWitt equation suitably generalized.[62] In addition we describe the relation between a particle Quantum Field Theory in the Megaverse and in a universe within the Megaverse. *We also describe the needed generalizations of Quantum Field Theory to adequately and consistently describe a universe quantum field's relation to its Megaverse counterpart. These generalizations are: the local definition of asymptotic particle states, Bogoliubov transformations between a universe quantum field and its Megaverse counterpart, Two Tier QFT to eliminate perturbation theory infinities, and Pseudoquantum Field Theory to accommodate Higgs vacuum expectation values and higher derivative interactions.*

7.1 Quantum Gravity in the Megaverse and its Universes

Since our universe is described by Quantum Gravity, and other universes are also, it is reasonable to assume the Megaverse is described by classical and Quantum Gravity. The source of Quantum Gravity in a universe is the mass-energy within the universe including particle masses, the Baryonic and Dark Baryonic gauge fields, the Two Tier Y^μ field, and interaction gauge fields.

The sources of classical Gravitation and Quantum Gravity in the Megaverse are analogously the "masses" of universe particles, the Baryonic and Dark Baryonic gauge fields, other gauge fields, the Ω-group gauge interactions, the mass-energy in the Megaverse outside of universes,[63] and the Megaverse Two Tier Y^μ field.

The surface of a universe in Quantum Gravity is fuzzy and determined by the quantum universe's wave function. This wave function is a solution of the universe Wheeler-Dewitt equation. The Wheeler-DeWitt equation assumes a single 4-dimensional universe with real-valued coordinates and metrics.

Since our theory requires complex coordinates and thus complex-valued metrics, we extended the Wheeler-DeWitt equation to complex metrics in I. Now we must define a Wheeler-DeWitt for the Megaverse's complex-valued coordinates and metric.

---

[62] Most of this chapter appeared in Blaha (2017b), Blaha (2015a) and earlier books by the author.

[63] We believe the Megaverse has a very low density mass-energy between universes.

7.2 Megaverse Wheeler-Dewitt Equation

The Megaverse Wheeler-DeWitt equation inside a universe (See I.) is

$$\Big(G_{ijkl}\Big\{\prod_x\sum_m\{[\delta^m{}_j\,\partial f_n/\partial x^i + \delta^m{}_i\,\partial f_m/\partial x^j\,](\partial^2 f_n/\partial x^{m2})\}^{-1}\partial/\partial x^m\Big\}\Big\{\prod_x\sum_m\{[\delta^m{}_k\,\partial f_n/\partial x^l + \delta^m{}_l\,\partial f_n/\partial x^k\,](\partial^2 f_n/\partial x^{m2})\}^{-1}\partial/\partial x^m\Big\} + \gamma^{\frac{1}{2}\,(3)}R + 2\lambda\gamma^{\frac{1}{2}\,(3)}\Big)\Psi({}^{(3)}G, L_F) = 0 \tag{7.2}$$

with universe coordinates x. In the Megaverse the Wheeler-DeWitt equation, in terms of Megaverse coordinates y_n is

$$0 = \Big(G_{ijkl}\Big\{\prod_x\sum_m\{[\delta^m{}_j\,\partial y_n/\partial x^i + \delta^m{}_i\,\partial y_m/\partial x^j\,](\partial^2 y_n/\partial x^{m2})\}^{-1}\partial/\partial x^m\Big\}\Big\{\prod_x\sum_m\{[\delta^m{}_k\,\partial y_n/\partial x^l + \delta^m{}_l\,\partial y_n/\partial x^k\,](\partial^2 y_n/\partial x^{m2})\}^{-1}\partial/\partial x^m\Big\} + \gamma^{\frac{1}{2}\,(D-1)}R + 2\lambda\,\gamma^{\frac{1}{2}\,(D-1)}\Big)\Psi({}^{(D-1)}G, L_F) \tag{7.3}$$

where the sums over n and m in each pair of {} are done independently. All references to the metric are expressed in terms of Megaverse quantiyties.

Due to the appearance of products over all coordinates, the Megaverse expression for $\delta/\delta\gamma_{ij}$ is independent of x, and of Megaverse coordinates, in accordance with the space-time independence of the original Wheeler-DeWitt equation.

The Megaverse form of the Wheeler-DeWitt equation also directly relates the metric of a universe to the Megaverse. Every universe has two sets of coordinates: universe coordinates, usually labeled x, embodying the curvature of the universe induced by gravitation, and Megaverse coordinates usually labeled y.

Outside of universes the Megaverse Wheeler-DeWitt equation becomes

$$(G_{ijkl}\,\delta/\delta\gamma_{ij}\,\delta/\delta\gamma_{kl} + \gamma^{\frac{1}{2}\,(D-1)}\,R + 2\lambda_M\,\gamma^{\frac{1}{2}\,(D-1)})\Psi({}^{(D-1)}\mathcal{G}) = 0 \tag{7.4}$$

with a cosmological constant, λ_M which is assumed to be present based on its presence in the universe case above and the uniformity of Nature. We assume a representation of the D-dimensional metric with a “spatial”sub-metric part γ_{ij00}. $\gamma^{\frac{1}{2}\,(D-1)}$ is the determinant of γ_{ik}.

Due to the products over all coordinates, the Megaverse expression for $\delta/\delta\gamma_{ik}$ is independent of Megaverse coordinates in analogy with the space-time independence of the original Wheeler-DeWitt equation. But the solutions of the Wheeler-DeWitt equations for a universe must be related to the solutions of the Wheeler-DeWitt equations of the Megaverse within the universe. This relationship must constrain the universe solutions at the boundary of a universe.

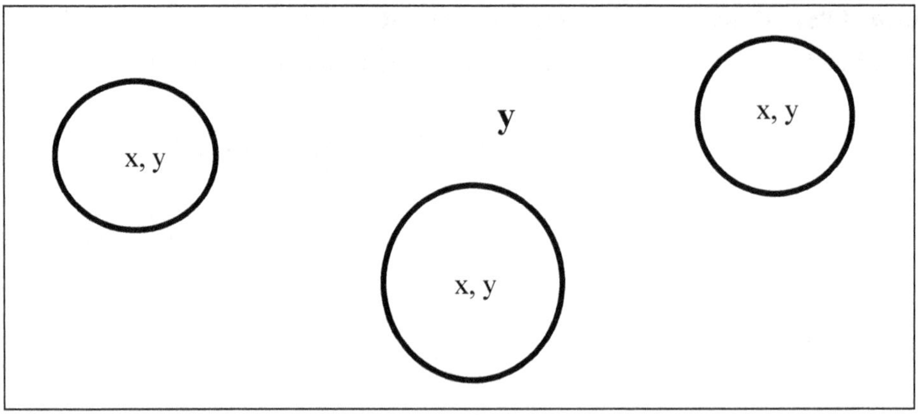

Figure 7.1. A symbolic view of part of the Megaverse with universes depicted as circles. Each universe has its own curvilinear coordinate 4-vector denoted x. The Megaverse coordinates are a D-vector y. A Wheeler-DeWitt equation applies within each universe and a comprehensive equation describes the entire Megaverse transitioning to each universe's wave function within its domain.

7.3 Megaverse Metric Functional Integral in a Universe

Corresponding to the functional derivative in eq. 7.3 is an explicit Megaverse form of the functional integral for a universe

$$\int D\gamma \equiv \int \prod_x \sum_{i,j} d\gamma_{ij}(x) = \int \prod_x \sum_{i,j,k} d(\partial f_k/\partial x^i \, \partial f_k/\partial x^j) \tag{7.5}$$

$$= \int \prod_x \sum_{i,j,k} [d(\partial f_k/\partial x^i)\, \partial f_k/\partial x^j + d(\partial f_k/\partial x^j)\, \partial f_k/\partial x^i]$$

$$= 2\int \prod_x \sum_{i,j,k} d(\partial f_k/\partial x^i)\, \partial f_k/\partial x^j$$

$$= 2\int \prod_x \sum_{i,j,k} dx(\partial^2 f_k/\partial x^{i2})\, \partial f_k/\partial x^j \equiv 2\int \delta f(x) \tag{7.6}$$

$$= 2\int \prod_x \sum_{i,j,k} dx(\partial^2 y_k/\partial x^{i2})\, \partial y_k/\partial x^j \equiv 2\int \delta f(x) \tag{7.7}$$

The third line is due to the summation over i and j. The factor of 2 can be absorbed in the normalization factor. Eqs. 7.5 – 7.7, like the Wheeler-DeWitt equation, are independent of the coordinates due to the product over coordinates x. Thus the solution of the Wheeler-DeWitt equation takes the form

$$\Psi({}^{(D-1)}\mathcal{G}, \mathcal{L}_F) = N \int \delta f(x) \exp(-I(g, \mathcal{L}_F)) \tag{7.8}$$

upon absorption of the factor of 2 into the normalization constant N

The Wheeler-DeWitt equation depends on the metric γ_{ij} which is a $(D-1)^2$ components construct with $D\cdot(D-1)/2$ independent components due to its symmetry.

7.4 Megaverse Wheeler-DeWitt Solutions

7.4.1 Tachyonic Solutions of Wheeler-DeWitt Equation

The original Wheeler-DeWitt equation, and its Megaverse equivalent, have a form that is similar in many respects to the Klein-Gordon equation. In particular, as DeWitt[64] noted, it resembles "a Klein-Gordon equation with $-\gamma^{1/2}\,{}^{(D-1)}R$ (our notation) playing the role of a mass-squared term. An important difference, however, is that ${}^{(D-1)}R$ can be either positive or negative, and hence the wave propagation of the state functional is not confined to time-like directions." DeWitt proceeds later in his paper to exclude consideration of negative "mass" squared terms.

In our view the negative "mass" squared cases represent tachyonic solutions of the Wheeler-DeWitt equation and should be considered as having the same validity as the positive mass squared solutions.

In a universe G_{ijkl} can be regarded as the contravariant metric of a 6-dimensional Riemannian manifold M with hyperbolic signature (-1, 1, 1, 1, 1, 1) with a time-like coordinate. Tachyonic solutions are part of the set of solutions of the Wheeler-DeWitt equation. In the Megaverse formulation the tachyonic solutions are indicators of tachyonic universes in the Megaverse.

If we use the Megaverse form of the Wheeler-DeWitt equation then the changes in time (dilations of γ_{ij} or equivalently dilations of y_n) can be viewed as specifying the overall motion of entire universes. We see now that we can have "normal" or tachyonic motion of universes.

Clearly we are building towards a particle view of universes.[65]

7.4.2 Problems in the Solutions of the Wheeler-DeWitt Equation

There are a number of problem areas associated with the original and our Megaverse Wheeler-DeWitt equations. DeWitt identified most of them in his paper, referenced below. While one could view these apparent problems as negatives, we will take the view that they are indicators of a deeper structure of universes below the Wheeler-DeWitt solutions just as the Dirac equation resolves analogous difficulties with the Klein-Gordon equation.

---

[64] DeWitt, B. S., Phys. Rev. **160**, 1113 (1987).p. 1124.

[65] The Reality groups of 4-dimensional universes, and of the Megaverse, play a role in the physical interpretation of Megaverse phenomena.

7.4.2.1 Negative Frequencies and Probabilities – Anti-Universes

DeWitt noted the existence of negative frequencies and negative probabilities associated with the solutions of the Wheeler-DeWitt equation. The rather close analogy to the Klein-Gordon equation in whose solutions similar issues appear is suggestive.

It appears that two general types of universes are embodied in the Wheeler-DeWitt equation. One type, which we call "normal", consists of universes like ours which have an excess of baryons (and consequently electrons to make an overall charge neutral universe). The other type of universe we will call *anti-universes*. These universes have an excess of anti-baryons (and positrons). We suggest such pairs can result from vacuum fluctuations in the Megaverse.

In addition we have "normal" universes and tachyonic universes whose motions in the Megaverse are analogous to the motions of fermions within a universe. In direct analogy with the Klein-Gordon equation the negative frequency and negative probability issues are thereby resolved.

One can speculate that the Wheeler-Dewitt equation, which is effectively second order in the "time" derivative, can be factored into first derivative equations – perhaps through the introduction of more degrees of freedom in a fashion similar to Dirac's introduction of spinors – to achieve first order equations in "time." Then we would have to face the issue of interpreting "Dirac-like" Wheeler-DeWitt equations.[66] We could again fall back on the rationale for introducing spinors for Standard Model fermions as in Blaha (2015a) – Asynchronous Logic – and suggest that universes embody a multi-valued Logic – although the concept of universes as logic values, at first glance, appears strange. We suggest that a universe's logic value is its spin.[67] Tied up with spin is "handedness." Our universe may be 'left-handed' as many phenomena within the universe seem to favor left-handedness. Therefore we can see a meaning for spin in universe internal properties. The "left handedness" of our universe suggests other universes with "right handedness" may exist.

We will be content, for the present, to assume both universes and anti-universes exist. An anti-universe will be presumed to be a universe in the Megaverse where anti-particles predominate. We will postulate that universes, and anti-universes like protons and anti-protons, are always bodies with extension and not point-like. Blaha (2013) and later chapters in this book provide an example of a Big Bang where a universe begins with extension and is not point-like.

Later we describe a proposed dynamics of universes and anti-universes based on Megaverse gauge fields.

7.5 Quantum Aspects of the Megaverse

The Megaverse contains quantized universes. Being quantized, the horizon (surface) of a universe is not precisely defined but undergoes presumably mild quantum smearing. In

[66] We will not consider factoring the Wheeler-DeWitt equation in this book.
[67] Fermionic universe spin is defined in chapter 3.

particular the surface of a quantum universe in the Megaverse must be defined, as particle positions are defined in quantum mechanics, by a wave function whose "square" at each Megaverse point is the probability of a part of the universe being there. Physically we would expect the probability to fall sharply shortly beyond the classical horizon (surface) of the universe.

So in a Megaverse with a low density of universes most of the Megaverse will have zero probability of a universe being present.

There are however four sources of quantum phenomena in the Megaverse: 1) gauge fields that provides interactions between universes as well as quantum fluctuations that create universes; 2) universe particles; 3) Megaverse mass-energy density; and 4) the Y^μ field that appears within universes and in the exterior Megaverse that quantizes the coordinates of each universe. A Y^μ fields existence in the Megaverse makes gauge field perturbation theory computations finite to all orders.

7.6 Quantum Field Theory in the Megaverse

The conventional form of Quantum Field Theory does not meet the requirements of a Physical theory in a Megaverse of more than four dimensions. Perturbation theory calculations will yield infinities – not predictions susceptible to physical interpretation. To remedy this flaw in Quantum Field Theory, and for other reasons cited below, we have introduced three enhancements to Quantum Field theory over the past fourteen years. We briefly describe these enhancements within the framework of the Megaverse in this section and refer the reader to previous books such as I (and references therein) that provide detailed descriptions.

7.6.1 Two Tier Quantum Field Theory in the Megaverse

Two Tier Quantum Field Theory,[68] which was based on a new method in the Calculus of Variations, uses two 'layers' of fields to introduce quantum coordinates. We shall consider this technique, which applies to all fields in I and here, for the specific case of a massless vector field $V^i(y)$ analogous to the electromagnetic field.

Since a field, quantized in D-dimensional conventional coordinates (D > 4), would lead to divergences in perturbation theory calculations we use D-dimensional quantum coordinates:

$$Y^i(y) = y^i + i\, Y_u^{\,i}(y)/M_u^{\,D/2} \tag{7.9}$$

where $Y_u^{\,i}(y)$ for i = 1, …, D is a D-dimensional free gauge field and M_u is a mass of the order of the Planck mass or greater. The $Y_u^{\,i}(y)$ term adds a quantum field to the D coordinates making them a set of quantum coordinates. Quantum coordinate derivatives are defined by

---

68 See Blaha (2005a), and Blaha (2002), for discussions of this new method to eliminate infinities in quantum field theory calculations.

$$\partial_i = \partial/\partial Y^i(y) = \partial/\partial(y^i - Y_u^{\ i}(y)/M_u^{\ D/2}) \tag{7.10}$$

The use of these coordinates to quantize particle fields leads to a completely finite perturbation theory. We applied them in I to create a finite fundamental theory of mater. We will apply them to fields in the Megaverse to achieve a finite theory of Megaverse dynamics for elementary particles and universe particles.

The second quantization of the vector gauge field, $V^i(y)$ is analogous to the second quantization of the electromagnetic field. The lagrangian density terms for the free $V^i(Y(y))$ fields is

$$\mathscr{L}_{Vu} = -\tfrac{1}{4}\, F_{Vu}^{\ \ ij}(Y(y))F_{Vuij}(Y(y)) \tag{7.11}$$

The lagrangian is

$$L_{Vu} = \int d^D y\, \mathscr{L}_{Vu}(Y(y)) \tag{7.12}$$

with

$$F_{Vuij} = \partial V_i(Y(y))/\partial Y^j(y) - \partial V_j(Y(y))/\partial Y^i(y) \tag{7.13}$$

where the values of i and j range from 1 to D in this section.

The equal time commutation relations, using the Dth coordinate as the time coordinate, are specified in the usual way:

$$[V^i(Y(\mathbf{y}, y^0)), V^j(Y(\mathbf{y}', y^0))] = [\pi^i(Y(\mathbf{y}, y^0)), \pi^j(Y(\mathbf{y}', y^0))] = 0 \tag{7.14}$$

$$[\pi_j(Y(\mathbf{y}, y^0)), V_k(Y(\mathbf{y}', y^0))] = -i\,\delta^{(D-1)tr}{}_{jk}(Y(\mathbf{y},0) - Y(\mathbf{y}',0)) \tag{7.15}$$

where

$$\pi_u^{\ k} = \partial\mathscr{L}_{Vu}(V(Y(y)))/\partial V_k'(Y(y)) \tag{7.16}$$

$$\pi_u^{\ D} = 0 \tag{7.17}$$

for k = 1, … , (D – 1), and

$$\delta^{(D-1)tr}{}_{jk}(\mathbf{y} - \mathbf{y}') = \int d^{(D-1)}k\; e^{i\,\mathbf{k}\cdot(Y(\mathbf{y},0) - Y(\mathbf{y}',0))}\,(\delta_{jk} - k_j k_k/\mathbf{k}^2)/(2\pi)^{D-1} \tag{7.18}$$

$$V_k'(Y(y)) = \partial V_k(Y(y))/\partial y^{1D} \tag{7.19}$$

for j, k = 1, 2, … , (D – 1).

If we choose the Coulomb gauge for $V_k(Y(y))$:

$$V^D(Y(y)) = 0$$

$$\partial V^j(Y(y))/\partial Y^j(y) = 0$$

for j = 1, 2, … , (D – 1) then (D – 2) degrees of freedom (polarizations) are present in the vector potential.[69] The Fourier expansion of the vector potential $V^i(Y(y))$ is:

$$V^i(Y(y)) = \int d^{(D-1)}k\, N_{0V}(k) \sum_{\lambda=1}^{D-2} \varepsilon^i(k, \lambda)[a_V(k,\lambda) :e^{-ik\cdot Y(y)}: + a_V^\dagger(k,\lambda) :e^{ik\cdot Y(y)}:] \quad (7.20)$$

for i = 1, … , (D – 2) where

$$N_{0V}(k) = [(2\pi)^{(D-1)} 2\omega_k]^{-1/2} \quad (7.21)$$

and (since the field is massless)

$$k^D = \omega_k = (\mathbf{k}^2)^{1/2} \quad (7.22)$$

where k^D is the energy, and where the $\varepsilon^i(k, \lambda)$ are the polarization unit vectors for λ = 1, … , (D – 2) and $k^\mu k_\mu = k^{D\,2} - \mathbf{k}^2 = 0$.

The commutation relations of the Fourier coefficient operators are:

$$[a_V(k,\lambda), a_V^\dagger(k',\lambda')] = \delta_{\lambda\lambda'}\delta^{D-1}(\mathbf{k} - \mathbf{k'}) \quad (7.23)$$
$$[a_V^\dagger(k,\lambda), a_V^\dagger(k',\lambda')] = [a_V(k,\lambda), a_V(k',\lambda')] = 0 \quad (7.24)$$

and the polarization vectors satisfy

$$\sum_{\lambda=1}^{D-2} \varepsilon_i(k, \lambda)\varepsilon_j(k, \lambda) = (\delta_{ij} - k_i k_j/\mathbf{k}^2) \quad (7.25)$$

The V^μ Feynman propagator is

$$iD_F^{trTT}(y_1 - y_2)_{jk} = \langle 0|T(V_j(Y(y_1))V_k(Y(y_2)))|0\rangle \quad (7.26)$$

$$= -\, ig_{jk} \int \frac{d^D k\, e^{-ik.(y_1 - y_2)}\, R(\mathbf{k}, y_1 - y_2)}{(2\pi)^{16}\, (k^2 + i\varepsilon)} \quad (7.27)$$

where g_{jk} is the D-dimensional Lorentz metric and where $R(\mathbf{k}, y_1 - y_2)$ is given by

$$R(\mathbf{k}, y_1 - y_2) = \exp[- k^i k^j \Delta_{Tij}(y_1 - y_2)/M_u^D] \quad (7.28)$$
$$= \exp\{ -k^2[A(v) + B(v)\cos^2\theta] / [(2\pi)^{D-2} M_u^4 z^2]\}$$

where k^2 is the sum of the squares of the D – 1 spatial components with

[69] Note we use the Coulomb gauge for Y(y) also.

$$z^\mu = y_1^\mu - y_2^\mu$$
$$z = |\mathbf{z}| = |\mathbf{y}_1 - \mathbf{y}_2|$$
$$k = |\mathbf{k}|$$
$$v = |z^0|/z$$
$$A(v) = (1 - v^2)^{-1} + .5v \ln[(v - 1)/(v + 1)]$$
$$B(v) = v^2(1 - v^2)^{-1} - 1.5v \ln[(v - 1)/(v + 1)]$$
$$\mathbf{k}\cdot\mathbf{z} = kz \cos\theta$$

and $|\mathbf{k}|$ denoting the length of a spatial (D – 1)-vector **k** while $|z^0|$ is the absolute value of $z^0 \equiv z^D$.

As eqs. 7.27 and 7.28 indicate, the Gaussian damping factor R(k, z) for *all* large spatial momentum k^j is the same for both the positive and negative frequency parts of the (Two Tier) V Feynman propagator. We are assuming the spatial momentum is real-valued in this discussion. It is also important to note that R(k, z) does not depend on $k^0 = k^D$ (in the V and Y_u Coulomb gauges) and thus the integration over k^0 proceeds in the usual way to produce time-ordered positive and negative frequency parts.

The Gaussian exponential factor in *all* spatial coordinates causes the Feynman propagator to be finite and, together with the Gaussian factor in universe particle propagators, causes all perturbation theory calculations when interactions are introduced to be finite as we have seen in I.

For small momentum much less than M_u then $R(\mathbf{k}, y_1 - y_2) \rightarrow 1$ and the Feynman propagator is the "normal" propagator of conventional D-dimensional quantum field theory. For large momentum the corresponding potential approaches r^{D-3} in contrast to the electromagnetic Coulomb potential r^{-1}. The V potential is highly non-singular at large energies.

Thus using Two Tier Quantum Field Theory we can perform perturbation theory caluculations that always yield a finite result.[70] This is not true if conventional Quantum Field is used.

7.6.2 Pseudoquantum Field Theory in the Megaverse

Pseudoquantum Field Theory which we developed in a series of books[71] also can be formulated in the Megaverse. Thus we can use it in the Megaverse to implement the Higgs Mechanism to generate particle masses and symmetry breaking.

In this section we generalize Pseudoquantum field theory to the Megaverse for a scalar field. It can be implemented for other particle fields in an analogous manner.

---

[70] In particular, the fermion triangle divergence (anomaly) does not occur in our Two Tier Quantum Field Theory of the fermion sector. Thus there is no requirement for axion-like particles in the Megaverse (or in universes) although the possible existence of this type of particle is not ruled out.

[71] See I for the discussion of the Pseudoquantum field theory formalism for Higgs particles in our Extended Standard Model. See chapter 20 of I, and earlier books, for a more detailed view than that presented here.

We will now Pseudoquantize a scalar particle field in the Megaverse that will become a Higgs particle with a non-zero vacuum expectation value.[72] We begin by defining two fields that correspond to the scalar particle:[73] $\varphi_1(x)$ and $\varphi_2(x)$ where x is a D-dimensional vector. These fields will be assumed to have the equal time commutators

$$[\varphi_a(x), \pi_b(y)] = i(1 - \delta_{ab})\delta^{(D-1)}(\mathbf{x} - \mathbf{y}) \qquad (7.29)$$
$$[\varphi_a(x), \varphi_b(y)] = 0$$
$$[\pi_a(x), \pi_b(y)] = 0$$

where δ_{ab} is the Kronecker δ and where $\pi_a(x)$ is the canonically conjugate momentum to $\varphi_a(x)$. The fields $\varphi_1(x)$ and $\pi_1(y)$ will be observable classical fields. The fields $\varphi_2(x)$ and $\pi_2(y)$ will not be observables so that $\varphi_1(x)$ and $\pi_1(y)$ can both be sharp on the set of physical states.

We now specify the lagrangian density for a scalar Megaverse Klein-Gordon particle:

$$\mathcal{L} = \partial\varphi_1/\partial x_\mu \partial\varphi_2/\partial x^\mu \qquad (7.30a)$$

with hamiltonian density

$$\mathcal{H} = \pi_1\ \pi_2 + \partial\varphi_1/\partial x_i \partial\varphi_2/\partial x^i \qquad (7.30b)$$

where μ labels 16-dimensional coordinates, i labels Megaverse spatial coordinates ((D – 1)-dimensional), and $\pi_1 = \partial\varphi_2/\partial t$ and $\pi_2 = \partial\varphi_1/\partial t$ with $t = x^D$. Eqs. 7.30 are without a potential or mass term.

The lagrangian and hamiltonian for a massive scalar particle are

$$\mathcal{L} = \partial\varphi_1/\partial x_\mu \partial\varphi_2/\partial x^\mu - m^2\ \varphi_1\varphi_2 \qquad (7.30c)$$

with hamiltonian density

$$\mathcal{H} = \pi_1\ \pi_2 + \partial\varphi_1/\partial x_i \partial\varphi_2/\partial x^i + m^2\ \varphi_1\varphi_2 \qquad (7.30d)$$

The massless fields can be fourier expanded in terms of creation and annihilation operators:

$$\varphi_i(\mathbf{x}, t) = \int d^{(D-1)}k\ [a_i(k)f_k(x) + a_i^\dagger(k)f_k^*(x)] \qquad (7.31)$$

for i = 1, 2 where

$$f_k(x) = N(k)e^{-ik\cdot x}$$

with N(k) being a normalization factor.

The creation and annihilation operators satisfy the commutation relations:

---

[72] Much of this chapter appears in Blaha (2016c), and earlier books, as well as in S. Blaha, Phys. Rev. **D17**, 994 (1978). The case of fermion Pseudoquantization is also discussed in S. Blaha, Il Nuovo Cimento **49A**, 35 (1979).

[73] The subscripts on the fields are not gauge symmetry indices but simply identifiers distinguishing the fields from each other.

$$[a_a(k), a_b^\dagger(k')] = (1 - \delta_{ab})\delta^{(D-1)}(\mathbf{k} - \mathbf{k'}) \tag{7.32}$$
$$[a_a(k), a_b(k')] = 0$$
$$[a_a^\dagger(k), a_b^\dagger(k')] = 0$$

for a, b = 1, 2.

In this formulation the defining properties of a physical state are:

$$\varphi_1(x)|\Phi, \Pi> = \Phi(x)|\Phi, \Pi> \tag{7.33}$$
$$\pi_1(x)|\Phi, \Pi> = \Pi(x)|\Phi, \Pi>$$

where $\Phi(x)$ and $\Pi(x)$ are sharp on the states and thus classical fields with

$$\Phi(\mathbf{x}, t) = \int d^{(D-1)}k\, [\alpha(k)f_k(x) + \alpha^*(k)f_k^*(x)] \tag{7.34}$$

and correspondingly for $\Pi(x)$.

To implement the mass generation mechanism we set Φ equal to a constant. We can define a set of states satisfying

$$a_1(k)|\alpha> = \alpha(k)|\alpha>$$
$$a_1^\dagger(k)|\alpha> = \alpha^*(k)|\alpha>$$

and correspondingly a set of coherent states

$$|\alpha> = C\exp\left\{\int d^3k\, [\alpha(k)a_2^\dagger(k) + \alpha^*(k)a_2(k)]\right\}|0> \tag{7.35}$$

where C is a normalization constant and where the vacuum state $|0>$ satisfies

$$a_1(k)|0> = a_1^\dagger(k)|0> = 0 \tag{7.36a}$$

$$a_2(k)|0> \neq 0 \qquad\qquad a_2^\dagger(k)|0> \neq 0 \tag{7.36b}$$

The dual vacuum state satisfies

$$<0|a_2(k) = <0|a_2^\dagger(k) = 0 \tag{7.37a}$$
$$<0|a_1(k) \neq 0 \qquad\qquad <0|a_1^\dagger(k) \neq 0 \tag{7.37b}$$

With this coherent state formalism, which gives purely classical fields and yet also has quantum fields through the use of φ_2 and its creation and annihilation operators, we now have the machinery to define a mass mechanism without the introduction of a potential whose origin can only be described as dubious.

For we can define a coherent state for some k as

$$|\Phi, \Pi> = C\exp\{[2N(k)]^{-1}\Phi[a_2^\dagger(k) + a_2(k)]\}|0> \quad (7.38)$$

where C is a normalization constant, that yields a non-zero vacuum expectation value:

$$\varphi_1(x)|\Phi, \Pi> = \Phi|\ \Phi, \Pi> \quad (7.39)$$

where Φ is a constant. Evaluating a fermion interaction term we find a mass term emerges[74]

$$\bar{\psi}(\varphi_1 + \varphi_2)\psi \rightarrow \bar{\psi}(\Phi + \varphi_2)\psi \quad (7.40)$$

It generates a mass for an interaction with a gauge field of the form

$$A^\mu(\varphi_1 + \varphi_2)^2 A_\mu \rightarrow A^\mu(\Phi + \varphi_2)^2 A_\mu \quad (7.41)$$

It also yields a quantum field theoretic interaction that would result in the production of ElectroWeak particles from these scalar fields. The production of Higgs particles that decay into ElectroWeak gauge particles has recently been found at CERN.

Thus our Pseudoquantum formalism is well-adapted to generate particle masses and symmetry breaking. Chapter 20 of I lists additional benefits of the use of Pseudoquatization. Section 7.6.3 below shows that we can define Quantum Field Theories that support quantization for arbitrary timelike directions with local definitions of asymptotic states.

7.6.3 Local Definition of Asymptotic Particle States

The local definition of particle states is a significant point of interest for the Megaverse. Given the need for a quantum formulation of particle theory, we must address the issue of particle field quantization in a universe vs. quantization in the Megaverse. A particle state of a field quantized in one coordinate system is, in general, a superposition of particle states if the particle field is quantized in a different coordinate system. This is true within a universe. It is also true if one quantizes a field within a universe's coordinate system, and also quantizes the field in a Megaverse coordinate system. Since a universe can be described in a universe coordinate system, or in Megaverse coordinates, the problem of field quantization in coordinate systems, and the interpretation of particle states, becomes more critical when Megaverse coordinates are brought into consideration.

Some years ago this author developed a formulation[75] of quantum field theory in which the particle interpretation of particle states was unambiguous: an n particle state in one field

[74] When matrix elements with a "vacuum state" such as eq. 3.10 are taken.

[75] S. Blaha, "The Local Definition of Asymptotic Particle States", IL Nuovo Cimento **49A**, 35 (1979). This paper is reprinted in Appendix A for the reader's convenience. Also the paper S. Blaha, "New Framework for Gauge Field

quantization's coordinate system was an n particle state for the field quantized in any other coordinate system. Thus the particle interpretation of states was independent of the coordinate system chosen for second quantization.

In this section we overview this method of quantization since it relates to the very real issue of the transition of particles between the Megaverse and a universe. If one envisions a particle (or a starship!) traveling between a universe and the Megaverse, then the fate of the particle (starship) after the transition is directly related to the possible(?) quantum field theoretic change of particle state(s).

The second quantization method described in this section is a generalization of the Pseudoquantization procedure described earlier in this section. The method was developed in the late 1970's by the author to provide a quantization procedure which supports a unique particle interpretation of states in arbitrary non-static space-times where no global timelike coordinate (Killing vector) exists. An N particle state in one quantization is an N particle state in other quantizations. Physical particle states of different quantizations are related by a unitary Bogoliubov transformation that preserves the particle number of the states. (The particle number operator commutes with the operator generating the unitary transformation.) See Appendix A for a detailed discussion.[76] We shall assume the reader has read Appendix A and extend the discussion to Megaverse quantization vs. universe quantization below.

We will consider the case of a scalar particle. Charged scalars, fermions and gauge fields are considered in Appendices A and B. These other cases are completely analogous.

7.6.3.1 Second Quantization and the Definition of Particle States in a Universe

Let us consider the case of a scalar particle that we second quantize in some fashion based on a timelike Killing vector

$$\varphi(x) = \sum_{\alpha} \chi_\alpha(x)A_\alpha + \chi_\alpha{}^*(x)A_\alpha{}^\dagger \tag{7.42}$$

where the $\chi_\alpha(x)$ are positive frequency with respect to a definition of positive frequency within a universe – following the notation of Appendix A.

Theories", IL Nuovo Cimento **49A**, 113 (1979) applies this quantization method to non-Abelian gauge theories. It is reprinted in Appendix B.

[76] The discussions in the papers of Apendices A and B were predicated on an assumption of one universe. The generalization to Megaverse-universe quantizations is straightforward. We note, as Appendix A points out, that differences in the quantizations of two relatively accelerating observers *do cause* different numbers of particles to be evident in corresponding states – both physically, and in the quantized theories particle states. The quantizations described here are for *one* observer using different coordinate systems. The case of *two* relatively accelerating observers is different as noted in Appendix A.

7.6.3.2 Second Quantization and the Definition of Particle States in the Megaverse

Consider now the case of the same scalar particle that we second quantize in the Megaverse based on a timelike Megaverse Killing vector

$$\varphi(y) = \sum_{\beta} \psi_\beta(y)b_\beta + \psi_\beta{}^*(y)b_\beta{}^\dagger \tag{7.43}$$

where the $\psi_\beta(x)$ are positive frequency with respect to a Megaverse definition of positive frequency.

7.6.3.3 Relation Between the Definitions of Quantized Fields and Particle States

Comparing eqs. 7.42 and 7.43 we note the difference in the definition of the coordinates used in the field expansions as well as the implicit difference in the definitions of positive frequency. Therefore, to relate the quantizations to each other *solely within a universe*, we must use the relation between Megaverse coordinates y and universe coordinates x:

$$y_i = f_i(x) \tag{7.44}$$

or, in vector form,

$$y = f(x) \tag{7.45}$$

for i = 1, 2, … , D. Thus

$$\varphi(f(x)) = \sum_{\beta} \psi_\beta(f(x))b_\beta + \psi_\beta{}^*(f(x))b_\beta{}^\dagger \tag{7.46}$$

Inverting the above equations to obtain the relation of the fourier coefficient operators we see:

$$A_\alpha = \sum_{\beta} [C_{\alpha\beta}\, b_\beta + C'_{\alpha\beta}\, b_\beta{}^\dagger] \tag{7.47}$$

where $C_{\alpha\beta}$ and $C'_{\alpha\beta}$ are c-number functions of α and β:

$$\begin{aligned} C_{\alpha\beta} &= (\chi_\alpha(x), \varphi(f(x)) \\ C'_{\alpha\beta} &= (\chi_\alpha{}^*(x), \varphi(f(x)) \end{aligned} \tag{7.48}$$

with eq 7.46 substituted for φ(f(x)) in the inner products, which are integrals over the universe coordinates x.

Eqs. 7.48 imply an N particle state in a universe will appear as a superposition of states of various numbers of particles in Megaverse coordinates IF THE STANDARD QUANTUM FIELD THEORY FORMULATION IS USED. A practical implication of this formalism is that a mouse in a universe is a superposition of protoplasm in the Megaverse – an unpleasant prospect for manned travel out of a universe into the Megaverse.

To REMEDY this situation – which we take to be unphysical[77] – we must reformulate quantum field theory in a manner similar to the Pseudoquantum formulation presented earlier.

The new formulation associates two fields with a particle in a manner very similar to that of Pseudoquantization field theory discussed earlier. The scalar particle case is discussed in Appendix A between eqs 6 – 31. The reader is directed to read that section.

The conclusions of that section, and the sections following it, in Appendix A are:

1. One can define corresponding particle states in a Megaverse quantization or in a universe quantization with the same number of particles.
2. The fourier coefficient operators of the two quantizations are related by Bogoliubov transformations and are unitarily equivalent.
3. The group of the local Bogoiubov transformations is an infinite tensor product of $SU_{1,1}$ groups.
4. The vacua of the particle are invariant under Bogoliubov transformations that relate the the Megaverse and the universe quantizations.
5. Unitarily equivalent perturbation theories of both the Megaverse and the universe quantizations can be defined.

The equations of Appendix A (and B) can be taken to apply to a universe quantization vs. a Megaverse quantization with the proviso that a map of Megaverse coordinates to universe coordinates must be used to calculate fourier coefficient operators such as in eqs. 7.46 – 7.48 above.

We thus have shown that our generalized Pseudoquantization formalism can be used to relate Megaverse quantum field theory in Megaverse coordinates to universe quantum field theory in universe coordinates.

7.7 Unruh Bath for Accelerating Starships?

The preceding discussion in section 7.6.3 eliminates the issue of a starship moving with large acceleration encountering a thermal bath of incoming particles and a 'hot' vacuum.

7.8 Integrity of Starships Passing into the Megaverse

The preceding discussion in section 7.6.3 also eliminates problems with the transition of starships from a universe into the Megaverse. Particles are particles – within and without the Megaverse.

---

[77] A mouse is a mouse whether in the universe or Megaverse since the transition between them does not involve a physical change in the mouse. Earlier we suggested the transition is smooth since all neighborhoods of every point of a universe contains an infinite number of exterior Megaverse points.

8. Megaverse Interactions, Gravitation, and Reality Group

8.1 Megaverse Reality Group vs. Our Universe's Reality Group

Our universe has a Reality group that is the source of the Standard Model interactions. This Reality group is necessitated by the Complex Lorentz group, which supports complex-valued coordinate transformations. Realitygroup transformations map complex-valued coordinate systems resulting from Complex Lorentz group transformations to real-valued coordinate systems. The universe Reality group has the form:

$$SU(3) \otimes SU(2) \otimes U(1) \otimes SU(2) \otimes U(1)$$

It is partitioned into factors due to the structure of the Complex Lorentz group.

The Megaverse also has complex coordinates in a D-dimensional space. It has a Complex D-dimensional Lorentz group and a Complex D-dimensional General Relativity.Their transformations yield complex-valued Megaverse coordinate systems. An acceptable physics in the Megaverse must, like in our universe, have a time coordinate and D – 1 spatial coordinates. Otherwise dynamics, as we understand it, would not be possible. With this choice of interpretation we obtain a form of coordinate transformations similar to those of our universe but in D dimensions.

Given the complex nature of the Megaverse Lorentz group and General Relativity we again face transformations that yield complex-valued coordinate systems. So again we must introduce a local Reality group that maps complex-valued coordinates to real-valued coordinates. In the Megaverse case we can choose the local unitary group to be U(D). Local U(D) can transform any Megaverse D-vector locally to a real-valued D-vector.

The choice of U(D) as the Reality group has the additional benefit that we can identify it as the Megaverse equivalent of the universe SU(3)⊗U(64) 'Ω-group, with universe gauge fields $A_\Omega^{\mu}(x)$. The Megaverse Reality group is represented by a singlet SU(3) representation and the D-dimensional fundamental representation of U(D.)

Then, as discussed in chapter 3, we can view the Megaverse Reality group as the extension (completion) of the definition of the 'Ω-group:

1) In the universe it rotates interaction fields,[78] and fermions, as shown in I. (See section 3.1)
2) In the Megaverse it plays the role of the coordinate Reality group as well as rotating the eight universe interaction fields and fermion fields.

The Megaverse Reality gauge fields generate an interaction which is the extension of the universe $A_\Omega{}^\mu(x)$ fields. The Megaverse fields have index values $\mu = 1, 2, \ldots, D$ with the Dth coordinate being the time coordinate. Their symmetry is the same as in the universe.

8.2 Megaverse 'Special Relativity'

Megaverse Special Relativity implements a Complex Lorentz group whose transformations L preserve the metric $g_{ij} = \text{diag}(-1, -1, \ldots, -1, +1)$:

$$L^T[g]L = [g] \quad (8.1)$$

where [g] is the matrix representation of the metric and ^{T} represents the transpose. The transformations are a direct generalization from the 4-dimensional case.

Since Megaverse coordinates apply within a universe as well, one can map the Lorentz transformations of the Megaverse to the Lorentz transformations of a universe.

8.3 Connection Between Universe Interactions and Megaverse Interactions

The eight vector interactions described in I are:

Strong	$\mathbf{A}_{SU(3)}{}^\mu(x)$
Weak	$\mathbf{W}^\mu(x)$
Electromagnetic	$\mathbf{A}_E{}^\mu(x)$
Dark Weak	$\mathbf{W}_D{}^\mu(x)$
Dark Electromagnetic	$\mathbf{A}_{DE}{}^\mu(x)$
Generation group	$\mathbf{U}^\mu(x)$
Layer group	$\mathbf{V}^\mu(x)$
General Relativistic Reality Group	$\mathbf{A}_S{}^\mu(x)$

In the Megaverse they generalize from 4-vectors in our universe to D-vectors in the Megaverse. The issue we now face is the connection between the interaction fields in a universe and in the

[78] Note that the spatial index of vector fields in the universe ranges from 0, … , 3 while in the Megaverse the index of each of the vector fields ranges from 1, … , D. This was discussed in I for the baryonic field for the case of D = 16. Each field in the universe with four index values is a special case of the Megaverse field with D index values.

Megaverse. Two cases present themselves according to our discussion of Megaverse surface tension.

8.3.1 No Leakage from the Universe

If there is an infinite surface pressure force on a universe, the interactions within the universe are insulated from the Megaverse. Thus universe interactions can be treated on an indepednt basis. Boundary conditions are not a concern

8.3.2 Leakage from the Universe into the Megaverse

If there is leakage from a universe due to a low surface pressure force, then there are boundary conditions, in principle, that relate each Megaverse field to its cohort within the universe. Our universe, being flat, has an infinite surface force preventing leakage.

8.4 Megaverse General Relativity

Megaverse General Relativity has the same form as that of a universe but indices range from 1, … , D in the Megaverse while thay have four values in a 4-dimensional universe. Obviously, the types of configurations and topologies will be more varied in a D-dimensional Megaverse.

In the case of our Megaverse, the very low density of mass-energy (sections 14.14 and 14.15) indicates the Megaverse is flat to very good approximation.

8.5 Interactions of Universes and the Megaverse

The particles of the Megaverse in Megaverse space (outside of universes) experience the gauge interactions of section 8.3 extended to D-vectors. They also experience the U(D) Reality group interaction and D-dimensional gravitation.

Universe particles (chapter 10) experience D-dimensional gravitation, multipole electromagnetic interactions, baryoninc and Dark baryonic interactions, and possibly leptonic number and Dark leptonic number interactions (if they are long-range).

Since the Higgs Mechanism for gauge fields and particle masses is uncertain in the Megaverse whether various gauge field interactions are long range is uncertain.

9. General Types of Universes within the Megaverse

In Blaha (2011c) we determined the dimensionality of our universe based on principles of Asynchronous Logic that suggested a 4-valued logic that could be embodied in a 4-dimensional spinor matrix formulation. This 4-dimensional spinor formulation led to a 4-dimensional space-time.

The requirement that the speed of light is the same in all inertial reference frames, and that transformations between reference frames in faster than light relative motion are physically allowed, led to the requirement of complex coordinates and the Complex Lorentz transformation group (as was found necessary in Axiomatic Quantum Field Theory studies.) The reality of all physical time and distance measurements led to the introduction of the Reality group that mapped complex quantities to real physical values. This chain of logic is in accord with Leibniz's minimax principle: nature uses the simplest means to create complex physical phenomena.

While the preceding paragraph applies very nicely to our universe, and presumably to other universes, the question of the existence of other universes within the Megaverse with different dimensions naturally arises. Stars and galaxies have many varieties. Why should all universes be of dimension four?

Having developed the fundamental nature of our universe from Logic (the only sure requirement of any physical theory) it seems reasonable to classify possible universes based on their fundamental logic. In Blaha (2011c) we developed matrix formulations for many-valued logics. Assuming no separate clock mechanism to synchronize parts of complex processes, we developed an $n \times n$ matrix formalism for n-valued logic.

Therefore we could develop a principal sequence of types of universes based on n-valued logic. We summarize the small n-valued cases in Table 9.1.

Fant (2005) points out that VLSI circuits with spatially separated parts, which require time synchronization of activity without clocks, need a 4-valued logic at minimum. Thus for a complex universe such as ours the minimum space-time dimensionality is 4. For a smaller number of dimensions the complexity of physical processes is much diminished as the many solvable models of low space-time dimensionality show: easily solved – not very complex phenomena!

Smaller dimensioned universes may well exist – but not with the richness of complexity that leads to our type of universe's phenomena such as life.

Larger dimension universes may well also exist. They would have an excess of phenomena that might preclude life as we know it, or engender new forms of life.

n-Valued Logic	Matrix Representation Size	Spinor Components	Space-Time Dimensionality[79]
1	1×1	1	1
2	2×2	2	2
3	3×3	2	3
4	4×4	4	4
5	5×5	4	5
6	6×6	8	6

Table 9.1. Space-time dimensionality and number of spinor components corresponding to various n-valued logics. It would seem that the minimal acceptable number of dimension is 4 if one is to have a physically acceptable universe as we understand it.

The general tendency of physical phenomena to be largely based on extrema suggests that 4-dimensional space-time based on Leibniz's minimax principle is the "logical" choice for all universes.The case of 6-dimensional universes also appears attractive for a number of reasons.

The above classification scheme for universes is based on logic. Another important consideration is size. It appears that universes can have differing sizes and in fact can also grow or diminish in size (expansion or contraction). We will consider the size issue in chapter 10 as the expansion/contraction of a universe due to a time-dependent mass. Other possible differentiating factors between universes will also be considered. Chapters 11 – 14 consider models for the Big Bang expansion and the current universe expansion.

[79] Weinberg (1995) p. 216 exhibits an equation that relates the number of components of a spinor to the dimension of its space-time.

10. The Particle Interpretation of Universes

The Wheeler-DeWitt equation, because of its similarity to the Klein-Gordon equation, has led to numerous proposals to view universes as particles.[80]

In this chapter[81] we will consider a possible particle interpretation of universes that, while consistent with the spirit of the Wheeler-DeWitt equation and the Megaverse, goes far beyond our current experimental knowledge, although some recent astronomical data tends to support it. It can only be justified in this century by its generality and simplicity. It just looks right.

The Wheeler-DeWitt equation specifies the internal dynamics of universes. The Megaverse, and its Baryonic, Dark Baryonic, Leptonic, Dark Leptonic gauge fields, and other gauge fields embody the dynamics of universes. There is also a D-dimensional Two Tier Y^μ field that eliminates potential infinities in perturbation theory calculations.

We view universes as extended particles in the D-dimensional Megaverse rather like hadrons in particle physics. As we did in the low energy days of particle physics we will first quantize universes as point-like particles. We then take account of their internal structure in interactions (between universes) using solutions of the Wheeler-DeWitt equation, spectral representations of vacuum expectation values, form factors, "deep inelastic" structure functions, and so on. The interaction between universes, and between a universe and an elementary baryon particle in another universe, can be similarly treated.

The sole interactions between universes are assumed to be gauge interactions and gravity. Gravity is present in the Megaverse outside of universes due to universe particles, Megaverse matter, and gauge fields. All interactions are cloaked with Y^μ fields through the use of Megaverse quantum coordinates similar to the quantum coordinates discussed in I.

10.1 The Hierarchy of the Cosmos

In our universe we have seen that natural phenomena form a hierarchy ranging from the simplest to the largest/most complex phenomena. One current view of the hierarchy of levels of physical phenomena is:

Elementary particles: leptons, quarks, gluons, gauge bosons, and Higgs particles

---

[80] Some suggestions of this interpretation are: DeWitt, B. S., Phys. Rev. **160**, 1113 (1967); Robles-Perez, S. J., arXiv:1212.4598 (2012); and references therein.

[81] This chapter is obtained from Blaha (2015a) with a change of dimension from 16 to D.

Hadrons: protons, neutrons, …
Molecules
Agglomerations of molecules
Macroscopic objects
Planets
Stars
Galaxies
Clusters of galaxies
Supergalaxies
The Universe

Each level generally has a set of "simplified" physical laws that describe its phenomena.[82] For example molecules have quantum mechanical laws and regularities that help to understand the phenomena at the molecular level.

Interestingly, while all phenomena at each level should be explainable by the laws at lower levels, and ultimately, all phenomena should be explainable at the level of elementary particles, connecting phenomena at different levels is often quite difficult and, in many cases, impossible.

Consequently, while we believe physical phenomena are ultimately reducible to the lowest level, the problem of relating phenomena at different levels is largely unresolved.

In this book we introduce new levels in the hierarchy of nature: the level of multiple universes, and the level of the all-encompassing Megaverse. In doing this, we seek to maintain what we know of our universe, as embodied in our Extended Standard Model and Quantum Gravity. We will now turn to a discussion of the universes level and a portrayal of universes as extended particles.

10.2 The Particle Interpretation of Extended Wheeler-DeWitt Equation Solutions

In earlier chapters we described features of the Wheeler-DeWitt equation that suggested that universes could be viewed as particles or anti-particles, or tachyons. The solutions of this equation are scalar wave functions on a manifold that are analogous to the solutions of the Klein-Gordon equation. The issues of negative probabilities, possible tachyonic solutions, and negative frequency solutions suggest a need for an appropriate particle interpretation of universes that can possibly resolve these problems.

Some physicists have taken the Wheeler-DeWitt equation as the starting point for a theory of a universe as a particle. The Wheeler-DeWitt equation describes the interior of a universe in a quantum framework.

[82] This point was often made by Nobelist Kenneth Wilson of Cornell and Ohio State Universities.

We will take a different approach using the Megaverse as the environment of universe particles that internally have Quantum Gravity, and externally have Megaverse Quantum Gravity.

We view a universe as an extended particle and begin by ignoring the detailed inner structure of universes. This approach is similar to the historical treatment of hadrons such as the proton as particles and developing a theory of them as fundamental particles using form factors, structure functions and so on to approximate their inner structure. Afterwards, as detailed data became available, the detailed investigation of the internal structure of hadrons using quark-parton models followed. We will pursue a similar theoretical development beginning with a theory of universes as extended particles in the D-dimension Megaverse. The internal structure of the particle universes will eventually be specified by the Wheeler-DeWitt equation expressed in Megaverse coordinates.

The two simplest choices for the nature of universes are "spin 0" *bosonic universes* and *fermionic universes* with odd half integer spin, s_M.[83] We will first consider the possibility of fermionic universes, and then briefly consider "spin 0" *bosonic* universes.

The first issue of fermionic universes (reminiscent of the discussions of spin in the 1920's) is the interpretation of spin states. We suggest that the upper $2^{D/2-1}$ components (with $2^{D/2-2}$ "spin up" and $2^{D/2-2}$ "spin down" states) of a fermionic universe wave function represent a left-handed universe with an excess number of baryons. The lower ($2^{D/2-1}$) components lead to right-handed anti-universes where there is an excess of anti-baryons. These associations are analogous to the interpretations of the Dirac electron wave function.[84]

The universe particle "spin up" and "spin down" states are distinguished by their interactions with gauge fields in a manner analogous to quantum electrodynamics.

10.3 "Free Field" Dynamics of Fermionic Universe Particles

We now consider universes as extended particles with an odd half integer spin – *fermionic universe particles* - in the D-dimensional Megaverse. In the Megaverse there are D 'Dirac' matrices with $2^{D/2}$ rows and $2^{D/2}$ columns that are the equivalent of the four Dirac matrices in four dimensions. We will denote these D matrices as $\gamma_M{}^i$ for i = 1, 2, … , D. They satisfy the anti-commutation relations:

$$\{\gamma_M{}^i, \gamma_M{}^j\} = 2\,\delta^{ij} \tag{10.1}$$

and thus form a Clifford algebra. We will choose y^D to be the time coordinate and thus make it pure imaginary with a Reality group transformation. (The D-dimensional Megaverse space is a complex Euclidean space.) Therefore γ^D will be hermitean ($(\gamma^D)^2 = 1$), and the γ^i matrices for i =

[83] Since the Megaverse is D-dimensional, the spin of fermionic universe particles was shown to be s_M in chapter 3.

[84] It is known that phenomena in our universe tend to be left-handed. If this feature of our universe's phenomena is also a property of the universe itself, then, since handedness is an attribute of spin, the treatment of a universe as having spin is not unreasonable.

1, ... , (D − 1) will be anti-hermitean with $(\gamma^i)^2 = -1$. The number of linearly independent matrices in D dimensions is 2^D.

The Megaverse metric is (by use of the Reality group) chosen to be

$$g^{ij} = -\delta^{ij}, \qquad g^{D,D} = 1 \tag{10.2}$$

for i, j = 1, 2, ... , (D − 1); and zero otherwise.

Except for the additional dimensions, fermion dynamics is quite similar to the 4-dimensional case. The free universe particle Dirac equation is

$$(i\gamma^i\partial_i + m)\psi(y) = 0 \tag{10.3}$$

summed over i = 1, 2, ... , D where the mass is assumed to be constant, and set by eq. 10.119 below. The derivative operator, is based on the use of quantum coordinates[85] (eq. 7.9)

$$Y^i(y) = y^i + i\, Y_u^{\,i}(y)/M_u^{\,D/2} \tag{10.4}$$

For i = 1, ..., D and is defined to be

$$\partial_i = \partial/\partial Y^i(y) = \partial/\partial(y_i - Y_{ui}(y)/M_u^{\,D/2}) \tag{10.5}$$

where *we assume* $M_u = M_c$ *with* M_c *being a very large mass scale of perhaps the order of the Planck mass.*

$Y_u^{\,i}$ is a D-dimensional Megaverse gauge field equivalent of the universe $Y^\mu(x)$ used in Two Tier renormalization (discussed in I):

$$Y^\mu(z) = z^\mu + i\, Y^\mu(z)/M_c^{\,2}$$

where $Y^\mu(z)$ is a free QED-like field. The $Y^i(y)$ quantum coordinates will be used in the Megaverse to eliminate potential divergences, in a manner similar to the case of our universe when universe particle interactions are introduced later.

10.3.1 Four Types of Fermionic Universe Particles

Assuming universe energies are real-valued,[86] there are four possible types of fermionic universe particles in the Megaverse that are analogous to the four species of fermion described in I (and Blaha (2010b)) for The Extended Standard Model. Two of these types are tachyonic. It

[85] Giving Two Tier renormalization. See chapter 7.

[86] The energy of universe particles need not be real-valued since universes can 'decay' – unlike elementary particles which are not subject to decay, by definition, since they are assumed to be *fundamental*. We choose to consider the case of universes with real-valued energies. The case of universes with complex-valued energies is a simple extension of the real-value cases considered here.

is important to note that DeWitt points out that the Wheeler-DeWitt equation has tachyonic solutions since the mass-like term dependent on ${}^{(3)}R$ can be positive or negative.[87] A negative mass is an indication of tachyonic behavior wherein the wave propagation of the state functional is not necessarily in time-like directions and is thus tachyonic.

Eq. 10.3 is a Dirac-type D-dimensional Dirac equation. There are three other general types of universe particle equations. (By assumption fermionic universes come in four species like fermions.) The derivation of the four types of universe particles is similar to the derivation of fermion types in the Extended Standard Model in 4-dimensional complex space-time given in Blaha (2010b). We will now consider the D-dimensional equivalent for universe particles in the Megaverse.

The general form of a pure D-dimensional complex Lorentz group[88] boost can be expressed in terms of a complex relative (D – 1)-velocity $\mathbf{v_c}$ between inertial reference frames. A D-dimensional coordinate boost has the form

$$\Lambda_C(\mathbf{v_c}) \equiv \Lambda_C(\omega, \mathbf{v_c}) = \exp[i\omega\hat{\mathbf{w}}\cdot\mathbf{K}] \tag{10.6}$$

where

$$\omega = (\omega_r^2 - \omega_i^2 + 2i\omega_r\omega_i\,\hat{\mathbf{u}}_r\cdot\hat{\mathbf{u}}_i)^{1/2} \tag{10.7}$$

and

$$\hat{\mathbf{w}} = (\omega_r\hat{\mathbf{u}}_r + i\omega_i\hat{\mathbf{u}}_i)/\omega \tag{10.8}$$

with all vectors being (D – 1)-dimensional spatial vectors. We define the real and imaginary unit vectors $\hat{\mathbf{u}}_r\cdot\hat{\mathbf{u}}_r = 1 = \hat{\mathbf{u}}_i\cdot\hat{\mathbf{u}}_i$ with the result

$$\hat{\mathbf{w}}\cdot\hat{\mathbf{w}} = 1 \tag{10.9}$$

The complex relative velocity is

$$\mathbf{v_c} = \hat{\mathbf{w}}\tanh(\omega) \tag{10.10}$$

The free dynamical equations of the four universe particle species will be generated by D-dimensional Lorentz boosts of the free Dirac equation of a universe particle at rest with the *requirement that the time variable* ($t = y^D$) *and energy are real in the resulting field equations.*[89] The procedure can most easily be performed in D-dimensional momentum space with the Megaverse coordinate space version of the generated equation determined from the momentum space version.

---

[87] DeWitt, B. S., Phys. Rev. **160**, 1113 (1967) p. 1124.

[88] The D-dimensional complex Lorentz group has similar features to the 4-dimensional complex Lorentz group. We shall only discuss it to the extent needed for our universe particle type's derivation. See Weinberg (1995) for the 4-dimensional Lorentz group – the D-dimensional Lorentz group generalizes directly from the features of the 4-dimensional Lorentz group.

[89] The D-dimensional "energy" must be real since it relates to the area of the universe – a real number.

10.3.1.1 Dirac-like Equation – Type I universe Particle

A positive energy plane wave solution of the Dirac equation eq. 10.3 for a universe particle at rest is

$$\psi(y) = \exp[-imt]w(0) \tag{10.11}$$

where we set $\partial_t = \partial/\partial y_D$ while temporarily ignoring the $Y_u^{\,i}(y)/M_u^{\,D/2}$ term. w(0) is a $2^{D/2}$ component spinor column vector. The solution ψ(y) satisfies the momentum space Dirac equation for a particle at rest:

$$(m\gamma^D - m)\psi(y) = 0 \tag{10.12}$$

The $2^{D/2} \times 2^{D/2}$ spinor matrix form of a D-dimensional Lorentz boost with a relative real velocity **v** of the Dirac matrices is[90]

$$S^{-1}(\Lambda(\mathbf{v}))\gamma^\nu S(\Lambda(\mathbf{v})) = \Lambda^\nu{}_\mu(\mathbf{v})\gamma^\mu \tag{10.13}$$

where $\Lambda^\nu{}_\mu(\mathbf{v})$ is a D-dimensional Lorentz boost. S(Λ(**v**)) has the form

$$S(\Lambda(\mathbf{v})) = \exp(-\omega\gamma^D\boldsymbol{\gamma}\cdot\mathbf{v}/(2|\mathbf{v}|))$$

$$= \cosh(\omega/2)I + \sinh(\omega/2)\gamma^D\boldsymbol{\gamma}\cdot\mathbf{p}/|\mathbf{p}| \tag{10.14}$$

with *real* ω = arctanh(|**v**|) and *real* **v.** |**p**| is the magnitude of the spatial (D – 1)-vector. Also

$$S^{-1}(\Lambda(\mathbf{v})) = \gamma^D S^\dagger(\Lambda(\mathbf{v}))\gamma^D = \exp(\omega\gamma^D\boldsymbol{\gamma}\cdot\mathbf{v}/(2|\mathbf{v}|))$$

$$= \cosh(\omega/2)I - \sinh(\omega/2)\gamma^D\boldsymbol{\gamma}\cdot\mathbf{p}/|\mathbf{p}| \tag{10.15}$$

If we now apply S(Λ(**v**)) to the momentum space Dirac equation of a particle at rest (eq. 10.12) we find

$$0 = S(\Lambda(\mathbf{v}))(m\gamma^D - m)\,\psi(y)$$
$$= [mS(\Lambda(\mathbf{v}))\gamma^D S^{-1}(\Lambda(\mathbf{v})) - m]S(\Lambda(\mathbf{v}))w(0)$$

A straightforward evaluation shows

$$mS(\Lambda(v))\gamma^D S^{-1}(\Lambda(v)) = g_{D\mu\nu}p^\mu\gamma^\nu = \not{p} \tag{10.16}$$

---

[90] **The indices ν and μ from this point in this chapter have values: 1, 2, … , D.**

where p is a momentum D-vector. In addition we define the D-dimension spinor ($2^{D/2}$ components)

$$S(\Lambda(v))w(0) = w(p) \tag{10.17}$$

which can be viewed as a "positive energy D Dirac spinor". The Dirac equation in momentum space has the familiar form:

$$(\not{p} - m)\exp[-ip{\cdot}y]w(p) = 0 \tag{10.18}$$

Eq. 10.18 implies the free, coordinate space Dirac equation:

$$(i\gamma^\mu\partial/\partial y^\mu - m)\psi(y) = 0 \tag{10.19}$$

We identify this equation as the dynamical equation of a type 1 universe particle. It corresponds to the free charged lepton elementary particle species Dirac equation in particle physics.

10.3.1.2 Complex Boosts

The form of the D-dimensional spinor boost transformation corresponding to the coordinate transformation eq. 10.6 is:

$$\begin{aligned} S_C(\omega, \mathbf{v_c}) \equiv S_C &= \exp(-\omega\gamma^D\boldsymbol{\gamma}\cdot\hat{\mathbf{w}}/2) \\ &= \cosh(\omega/2)I + \sinh(\omega/2)\gamma^D\boldsymbol{\gamma}\cdot\hat{\mathbf{w}} \end{aligned} \tag{10.20}$$

with *complex* $\mathbf{v_c}$ and $\hat{\mathbf{w}}$ defined by eqs. 10.10 and 10.8 respectively. The inverse transformation is

$$\begin{aligned} S_C^{-1}(\omega, \mathbf{v_c}) &= \exp(\omega\gamma^D\boldsymbol{\gamma}\cdot\hat{\mathbf{w}}/2) \\ &= \cosh(\omega/2)I - \sinh(\omega/2)\gamma^D\boldsymbol{\gamma}\cdot\hat{\mathbf{w}} \end{aligned} \tag{10.21}$$

Note that S_C is not unitary just as in the 4-dimensional case.

We now apply a spinor boost to the Dirac equation for a particle at rest in this more general case of complex ω and $\hat{\mathbf{w}}$.

$$\begin{aligned} 0 &= S_C(\omega, \mathbf{v_c}))(m\gamma^D - m)\exp[-imt]w(0) \\ &= [mS_C\gamma^D S_C^{-1} - m]\exp[-imt]S_C w(0) \end{aligned} \tag{10.22}$$

where $S_C = S_C(\omega, \mathbf{v_c})$. After some algebra we find

$$mS_C\gamma^D S_C^{-1} = m[\cosh(\omega)\gamma^D - \sinh(\omega)\boldsymbol{\gamma}\cdot\hat{\mathbf{w}}] \tag{10.23}$$

We will use these *complex* boosts to generate the other species' Dirac-like equations.

10.3.1.3 Tachyon Universe particle Dirac Equation

The development of the complex spinor boost transformation (subsection 10.3.1.2 above) leads to two possible forms of the tachyon Dirac-like equation. One form will lead to a lagrangian dynamics for left-handed universe particles. The other form leads to a lagrangian dynamics for right-handed universe particles.

10.3.1.4 Type IIa Case: Left-Handed Tachyonic Universe Particles

If the real and imaginary relative vectors parts of $\hat{\mathbf{w}}$, namely $\hat{\mathbf{u}}_r$ and $\hat{\mathbf{u}}_i$, are parallel, then $\hat{\mathbf{u}}_r\cdot\hat{\mathbf{u}}_i = 1$ and

$$\omega = \omega_r + i\omega_i \tag{10.24}$$

Eqs. 10.23 and 10.24 then imply

$$mS_C\gamma^D S_C^{-1} = m[\cosh(\omega_r)\cos(\omega_i) + i\sinh(\omega_r)\sin(\omega_i)]\gamma^D - \\ - m[\sinh(\omega_r)\cos(\omega_i) + i\cosh(\omega_r)\sin(\omega_i)]\boldsymbol{\gamma}\cdot\hat{\mathbf{u}}_r \tag{10.25}$$

or

$$mS_C\gamma^D S_C^{-1} = \cos(\omega_i)\gamma\cdot p_r + i\sin(\omega_i)\gamma\cdot p_i \tag{10.26}$$

where

$$p_r^0 = m\cosh(\omega_r) \qquad p_i^0 = m\sinh(\omega_r) \tag{10.27}$$

and

$$\mathbf{p}_r = m\hat{\mathbf{u}}_r\sinh(\omega_r) \qquad \mathbf{p}_i = m\hat{\mathbf{u}}_r\cosh(\omega_r) \tag{10.28}$$

If $\omega_i = 0$, then we recover the momentum space Dirac-like equation. If $\omega_i = \pi/2$, then we obtain the left-handed momentum space tachyon equation:

$$mS_C\gamma^D S_C^{-1} = i\gamma\cdot p_i \tag{10.29}$$

and the tachyon energy and momentum expressions

$$\mathbf{p} = m\mathbf{v}\gamma_s \qquad E = m\gamma_s \tag{10.30}$$

where $\sinh(\omega) = \gamma_s = (\beta^2 - 1)^{-1/2}$ with $\beta = v/c > 1$. v is the absolute value of the (D – 1) component spatial velocity. Also

$$S_C w(0) = w_C(p) \tag{10.31}$$

is a tachyon spinor.

The momentum space tachyonic Dirac-like equation is

$$(i\not{p} - m)\exp[-ip\cdot y]w_T(p) = 0 \tag{10.32}$$

where $p\cdot y = p^D y^D - \mathbf{p}\cdot\mathbf{y}$ after performing a corresponding boost in the exponential factor. If we apply $i\not{p}$ to eq. 10.32 we find the tachyon mass condition is satisfied

$$-E^2 + \mathbf{p}^2 = m^2 \tag{10.33}$$

Transforming back to coordinate space we obtain the "left-handed" *tachyonic Dirac-like equation*:

$$(\gamma^\mu \partial/\partial y^\mu - m)\psi_T(y) = 0 \tag{10.34}$$

10.3.1.5 Type IIb Case: Right-Handed Tachyonic Universe Particles

If the real and imaginary relative vectors parts of $\hat{\mathbf{w}}$, $\hat{\mathbf{u}}_r$ and $\hat{\mathbf{u}}_i$, are anti-parallel $\hat{\mathbf{u}}_r = -\hat{\mathbf{u}}_i$, then $\hat{\mathbf{u}}_r\cdot\hat{\mathbf{u}}_i = -1$ and

$$\omega = \omega_r - i\omega_i \tag{10.35}$$

then

$$mS_C\gamma^D S_C^{-1} = m[\cosh(\omega_r)\cos(\omega_i) - i\sinh(\omega_r)\sin(\omega_i)]\gamma^D - \\ - m[\sinh(\omega_r)\cos(\omega_i) - i\cosh(\omega_r)\sin(\omega_i)]\boldsymbol{\gamma}\cdot\hat{\mathbf{u}}_r \tag{10.36}$$

or

$$mS_C\gamma^D S_C^{-1} = \cos(\omega_i)\gamma\cdot p_r - i\sin(\omega_i)\gamma\cdot p_i \tag{10.37}$$

where

$$p_r^{\mathbf{D}} = m\cosh(\omega_r) \qquad p_i^{\mathbf{D}} = m\sinh(\omega_r) \tag{10.38}$$

and

$$\mathbf{p}_r = m\hat{\mathbf{u}}_r\sinh(\omega_r) \quad \mathbf{p}_i = m\hat{\mathbf{u}}_r\cosh(\omega_r) \tag{10.39}$$

If $\omega_i = \pi/2$, then we obtain the right-handed momentum space tachyon equation.[91]

$$(-\gamma^\mu \partial/\partial y^\mu - m)\psi_T(y) = 0 \tag{10.40}$$

10.3.1.6 Type III Case: "Up-Quark-like" Universe Particles

There are two other cases where we can obtain fermion dynamical equations with a *real* time variable and real energy.[92] In one case we set $\hat{\mathbf{u}}_r\cdot\hat{\mathbf{u}}_i = 0$ and have a real ω.

[91] We note that $\gamma_s = (\beta^2 - 1)^{-1/2}$, *if expressed in terms of ω, has a branch cut extending from* <-∞, +∞> *in the complex ω plane. Thus values of ω with positive imaginary parts are physically different from values of ω with negative imaginary parts.*

If the real and imaginary relative vectors parts of $\hat{\mathbf{w}}$, namely $\hat{\mathbf{u}}_r$ and $\hat{\mathbf{u}}_i$, are perpendicular, $\hat{\mathbf{u}}_r \cdot \hat{\mathbf{u}}_i = 0$, then

$$\omega = (\omega_r^2 - \omega_i^2)^{1/2} \qquad (10.41)$$

Thus ω is either pure real ($\omega_r \geq \omega_i$) or pure imaginary ($\omega_r < \omega_i$).

The momentum space equation generated by the corresponding spinor boost is

$$\{m \cosh(\omega)\gamma^D - m \sinh(\omega)\gamma\cdot(\omega_r\hat{\mathbf{u}}_r + i\omega_i\hat{\mathbf{u}}_i)/\omega - m\} \exp[-imt]w_c(p) = 0 \qquad (10.42)$$

Defining the momentum 4-vector

$$p = (p^D, \mathbf{p}) \qquad (10.43)$$

where

$$p^D = m \cosh(\omega) \qquad \mathbf{p} = \mathbf{p}_r + i\mathbf{p}_i \qquad (10.44)$$

with

$$\mathbf{p}_r = m\omega_r\hat{\mathbf{u}}_r \sinh(\omega)/\omega \qquad \mathbf{p}_i = m\omega_i\hat{\mathbf{u}}_i \sinh(\omega)/\omega \qquad (10.45)$$

$$\mathbf{p}_r \cdot \mathbf{p}_i = 0 \qquad (10.46)$$

then we obtain a positive energy Dirac-like equation

$$[p\cdot\gamma - m]\exp[-imt]w_c(p) = 0$$

or

$$[p^D\gamma^D - (\mathbf{p}_r + i\mathbf{p}_i)\cdot\gamma - m]\exp[-ip\cdot y]w_c(p) = 0 \qquad (10.47)$$

with a complex 3-momentum $\mathbf{p}$ and the 4-momentum mass shell condition:

$$p^2 = (p^D)^2 - \mathbf{p}_r\cdot\mathbf{p}_r + \mathbf{p}_i\cdot\mathbf{p}_i = m^2 \qquad (10.48)$$

Note

$$|\mathbf{v}_c| = |\mathbf{p}|/p^D = [(\mathbf{p}_r + i\mathbf{p}_i)\cdot(\mathbf{p}_r + i\mathbf{p}_i)]^{1/2}/p^D = \tanh(\omega) \qquad (10.49)$$

and so the Lorentz factor is

$$\gamma = \cosh(\omega) \qquad (10.50)$$

Eq. 10.47 is the momentum space equivalent of the wave equation[93]

---

[92] The requirement of a real energy for a universe is not strict. For a fundamental free particle the energy must be real or the particle would be subject to decay – contrary to its assumed fundamental nature. Universes can 'decay' to 'smaller' universes. Therefore the requirement for real energy can be violated. Nevertheless the requirement for real energy is appealing since it leads to four species of universes strengthening the analogy of universes to elementary particles.

[93] The gradient operators ∇_r and ∇_i are 15-dimensional spatial gradient operators.

$$[i\gamma^D\partial/\partial t + i\boldsymbol{\gamma}\cdot(\nabla_r + i\nabla_i) - m]\psi_u(t, \mathbf{y}_r, \mathbf{y}_i) = 0 \quad (10.51)$$

where $\mathbf{y} = \mathbf{y}_r - i\mathbf{y}_i$, and where the grad operators ∇_r and ∇_i are with respect to $\mathbf{y}_r$ and $\mathbf{y}_i$ respectively. Since $\hat{\mathbf{u}}_r\cdot\hat{\mathbf{u}}_i = 0$ we see that there is a subsidiary condition on the wave function

$$\nabla_r\cdot\nabla_i\, \psi_u(t, \mathbf{y}_r, \mathbf{y}_i) = 0 \quad (10.52)$$

We note eq. 10.52 can be put into covariant form as the difference of two vectors squared (which is a real D-dimensional Lorentz group invariant):

$$[\gamma^D\partial/\partial t + i\boldsymbol{\gamma}\cdot(\nabla_r + i\nabla_i)]^2 - [\gamma^D\partial/\partial t + i\boldsymbol{\gamma}\cdot(\nabla_r - i\nabla_i)]^2 = 4\nabla_r\cdot\nabla_i.$$

We identify eq. 10.51 as the dynamical equation of an "up-quark-like" universe particle.

10.3.1.7 Type IVa Case: Left-Handed "Down-Quark-like" Tachyonic Universe Particles

In this case we set $\hat{\mathbf{u}}_r\cdot\hat{\mathbf{u}}_i = 0$. Then by eq. 10.7

$$\omega = (\omega_r^2 - \omega_i^2)^{1/2}$$

Thus ω again starts out either pure real (if $\omega_r \geq \omega_i$) or pure imaginary (if $\omega_r < \omega_i$). In this case we also choose ω real, and then change ω to

$$\omega = (\omega_r^2 - \omega_i^2)^{1/2} \rightarrow \omega' = (\omega_r^2 - \omega_i^2)^{1/2} + i\pi/2 = \omega + i\pi/2$$

by adding $i\pi/2$ to ω since ω is a free parameter. We then proceed as we did in the prior tachyon case.[94]. The resulting Lorentz boost

$$\Lambda_C = \exp[i((\omega_r^2 - \omega_i^2)^{1/2} + i\pi/2)(\omega_r\hat{\mathbf{u}}_r + i\omega_i\hat{\mathbf{u}}_i)\cdot\mathbf{K}/\omega] \quad (10.53)$$

becomes a left-handed "quark-like" boost. The tachyon dynamical equation is[95]

$$[\gamma^D\partial/\partial t + \boldsymbol{\gamma}\cdot(\nabla_r + i\nabla_i) - m]\psi_d(\mathbf{y}) = 0 \quad (10.54)$$

with the constraint equation

[94] Here again the choice of ω in eq. 10.53 leads to a "left-handed" universe particle while the choice $\omega' = \omega - i\pi/2$ leads to a right-handed one.

[95] The gradient operators ∇_r and ∇_i are (D – 1)-dimensional spatial gradient operators.

$$\nabla_r \cdot \nabla_i \, \psi_d(t, \mathbf{y}_r, \mathbf{y}_i) = 0 \qquad (10.55)$$

We will call the universe particles satisfying eqs. 10.54 and 10.55 left-handed *tachyonic quark-like universe particles.*

10.3.1.8 Type IVb Case: Right-Handed Down-Quark-like Tachyonic Universe Particles

In this case we set $\hat{\mathbf{u}}_r \cdot \hat{\mathbf{u}}_i = 0$. Then by eq. 10.7

$$\omega = (\omega_r^2 - \omega_i^2)^{1/2}$$

Thus ω again starts out either pure real (if $\omega_r \geq \omega_i$) or pure imaginary (if $\omega_r < \omega_i$). In this case we also choose ω real, and then change ω to

$$\omega = (\omega_r^2 - \omega_i^2)^{1/2} \rightarrow \omega' = (\omega_r^2 - \omega_i^2)^{1/2} - i\pi/2 = \omega - i\pi/2$$

since ω is a free parameter and proceed as we did in the prior case. The resulting Lorentz boost

$$\Lambda_C = \exp[i((\omega_r^2 - \omega_i^2)^{1/2} - i\pi/2)(\omega_r \hat{\mathbf{u}}_r + i\omega_i \hat{\mathbf{u}}_i) \cdot \mathbf{K}/\omega] \qquad (10.56)$$

becomes a right-handed quark-like boost. The resulting tachyon dynamical equation is

$$[-\gamma^D \partial/\partial t - \boldsymbol{\gamma} \cdot (\nabla_r + i\nabla_i) - m]\psi_d(y) = 0 \qquad (10.57)$$

with the constraint equation

$$\nabla_r \cdot \nabla_i \, \psi_d(t, \mathbf{y}_r, \mathbf{y}_i) = 0 \qquad (10.58)$$

We will call the universe particles satisfying eqs. 10.57 and 10.58 right-handed *tachyonic quark-like universe particles.*

10.3.2 Lagrangians

In this section we will develop a lagrangian formalism for each of the four types of fermionic universe particles noting that a tachyonic universe particles have two forms: left-handed and right-handed (discussed later in section 10.3.5).

The various types of universe particles described in section 10.3.1 correspond to universes with differing internal characteristics and motion in the Megaverse. The equations are all free field equations. Internal potentials and interactions must be introduced in these equations to complete the universe dynamical equations. A connection to the Wheeler-DeWitt description of their internal quantum structure also remains to be established (section 10.3.6).

In defining the lagrangians for the four fermionic universe types that yield their dynamical equations in a canonical manner, we require the conventional quantum field theory feature that the hamiltonian derived from the lagrangian is hermitean. We will develop a separate lagrangian for each type.

10.3.2.1 Type I Universe Particle Lagrangian

The Universe particle Dirac equation lagrangian is

$$\mathcal{L}_u = \bar{\psi}(i\gamma^\mu\partial/\partial y^\mu - m)\psi(y) \tag{10.59}$$

where

$$\bar{\psi} = \psi^\dagger\gamma^D$$

and $\psi^\dagger$ is the hermitean conjugate of ψ.

10.3.2.2 Type II Tachyon Universe Particle Lagrangian

This lagrangian includes both left-handed and right-handed cases. It can be separated into lagrangian terms for each case using parity projection operators.

$$\mathcal{L}_{uT} = \psi_T{}^S(\gamma^\mu\partial/\partial y^\mu - m)\psi_T(y) \tag{10.60}$$

where

$$\psi_T{}^S = \psi_T{}^\dagger\, i\gamma^D\gamma^5 \tag{10.61}$$

with γ^5 being the D-dimensional equivalent for γ^5 in 4 dimensions. The peculiar form of the tachyon universe lagrangian is necessitated by the hermiticity of the hamiltonian calculated from it.

10.3.2.3 Type III "Up-Quark-like" Universe Particle Lagrangian

The lagrangian density of a free "up-quark-like" universe particle is

$$\mathcal{L}_u = \bar{\psi}_u(i\gamma^\mu D_\mu - m)\psi_u(y) \tag{10.62}$$

where $\bar{\psi}_u = \psi_u{}^\dagger\gamma^D$ and

$$\psi_u{}^\dagger = [\psi_u(\mathbf{y}_r, \mathbf{y}_i)]^\dagger\,|_{\mathbf{y}_i = -\mathbf{y}_i} \tag{10.63}$$

$$D_D = \partial/\partial y^D$$
$$D_k = \partial/\partial y_r{}^k + i\,\partial/\partial y_i{}^k \tag{10.64}$$

for k = 1, 2, … , (D – 1). The action

$$I = \int d^{(D-1)}y \mathcal{L}_u \tag{10.65}$$

It is easy to show that this action is also real.

10.3.2.4 Type IV "Down-Quark-like" Tachyon Universe Particle Lagrangian

The lagrangian density of a free "down-quark-like" universe particle is

$$\mathcal{L}_d = \psi_d^C(y)(\gamma^D\partial/\partial t + \boldsymbol{\gamma}\cdot(\nabla_r + i\nabla_i) - m)\psi_d(y) \tag{10.66}$$

where

$$\psi_d^C(y) = [\psi_d(y)]^\dagger|_{\mathbf{y}_i = -\mathbf{y}_i}\, i\gamma^D\gamma^5 \tag{10.67}$$

In words, eq. 10.67 states: take the hermitean conjugate of $\psi_d(y)$; change $\mathbf{y}_i$ to $-\mathbf{y}_i$; and then post-multiply by the indicated factors.

The action is

$$I = \int d^{(D-1)}y \mathcal{L}_d \tag{10.68}$$

The action is real. The lagrangian can also be separated into left-handed and right-handed parts using projection operators.

10.3.3 Form of The Megaverse Quantum Coordinates Gauge Field

The discussions of sections 10.3.1 and 10.3.2 assumed the coordinates were Megaverse coordinates and their derivatives. Prior to those discussions we indicated we would use quantum coordinates in the Megaverse of the form[96]

$$Y^i(y) = y^i + i\, Y_u^i(y)/M_u^{D/2} \tag{10.4}$$

and their derivatives

$$\partial_i = \partial/\partial Y^i(y) = \partial/\partial(y^i - Y_u^i(y)/M_u^{D/2}) \tag{10.5}$$

for i = 1, 2, … , D to eliminate divergences in quantum field theory. The subscript "u" signifies universes. The mass constant for the Megaverse, M_u, may be the same as the mass constant M_c appearing in the Two Tier mechanism for our universe. (See chapter 7 for a discussion of eliminating infinities with this mechanism.)

[96] The denominator $M_u^{D/2}$ is necessitated by the dimension of $Y_u^i(y)$ which is $[m]^{D/2-1}$. Eqs. 10.78 and 10.81 below imply this conclusion. See eq. 7.9.

In this section we define the gauge fields $Y_u^i(y)$ of the Megaverse.[97] They are similar to the $Y^\mu(y)$ fields of our New Standard Model.[98] The $Y_u(y)$ D-dimensional vector gauge field, in the absence of external sources, will be defined in a D-dimensional Coulomb gauge:

$$Y_u^D(y) = 0 \qquad (10.69)$$
$$\partial Y_u^j(y)/\partial y^j = 0$$

where the sum over j is over the D – 1 spatial y coordinates. We follow a procedure similar to Blaha (2003) but for D-dimensional space. The lagrangian density for the free $Y_u^j(y)$ fields is

$$\mathscr{L}_u = -\tfrac{1}{4} F_u^{\mu\nu} F_{u\mu\nu} \qquad (10.70)$$

and the lagrangian is

$$L_u = \int d^{(D-1)}y\, \mathscr{L}_u \qquad (10.71a)$$

with

$$F_{u\mu\nu} = \partial Y_{u\mu}/\partial y^\nu - \partial Y_{u\nu}/\partial y^\mu \qquad (10.71b)$$

The equal time commutation relations, derived in the usual way, are:

$$[Y_u^\mu(\mathbf{y}, y^0), Y_u^\nu(\mathbf{y}', y^0)] = [\pi_u^\mu(\mathbf{y}, y^0), \pi_u^\nu(\mathbf{y}', y^0)] = 0 \qquad (10.72)$$
$$[\pi_u^j(\mathbf{y}, y^0), Y_{uk}(\mathbf{y}', y^0)] = -i\,\delta^{(D-1)tr}{}_{jk}(\mathbf{y} - \mathbf{y}') \qquad (10.73)$$

for μ, ν, j, k = 1, 2, … , (D – 1) where

$$\pi_u^k = \partial\mathscr{L}_u/\partial Y_{uk}' \qquad (10.74)$$
$$\pi_u^0 = 0 \qquad (10.75)$$

and

$$\delta^{tr}{}_{jk}(\mathbf{y} - \mathbf{y}') = \int d^{(D-1)}k\, e^{i\,\mathbf{k}\bullet(\mathbf{y}-\mathbf{y}')}\,(\delta_{jk} - k_j k_k/\mathbf{k}^2)/(2\pi)^{D-1} \qquad (10.76)$$

$$Y_{uk}' = \partial Y_{uk}/\partial y^D \qquad (10.77)$$

The Coulomb gauge indicates D – 2 degrees of freedom are present in the vector potential. The Fourier expansion of the vector potential is:

$$Y_u^i(y) = \int d^{(D-1)}k\, N_0(k) \sum_{\lambda=1}^{D-2} \varepsilon^i(k, \lambda)[a(k,\lambda)\, e^{-ik\cdot y} + a^\dagger(k,\lambda)\, e^{ik\cdot y}] \qquad (10.78)$$

where

[97] This choice implies that the Megaverse Y mass $M_u = M_C$, its universe mass.
[98] See Blaha (2005a) for details.

$$N_0(k) = [(2\pi)^{(D-1)} 2\omega_k]^{-½} \tag{10.79}$$

and (since the field is massless)

$$k^D = \omega_k = (\mathbf{k}^2)^{½} \tag{10.80}$$

where k^D is the energy, and where the $\varepsilon^i(k, \lambda)$ are the polarization unit vectors for $\lambda = 1, \ldots, (D - 2)$ and $k^\mu k_\mu = k^{D\,2} - \mathbf{k}^2 = 0$.

The commutation relations of the Fourier coefficient operators are:

$$[a(k,\lambda), a^\dagger(k',\lambda')] = \delta_{\lambda\lambda'}\delta^{(D-1)}(\mathbf{k} - \mathbf{k'}) \tag{10.81}$$

$$[a^\dagger(k,\lambda), a^\dagger(k',\lambda')] = [a(k,\lambda), a(k',\lambda')] = 0 \tag{10.82}$$

and the polarization vectors satisfy

$$\sum_{\lambda=1}^{D-2} \varepsilon_i(k, \lambda)\varepsilon_j(k, \lambda) = (\delta_{ij} - k_i k_j/\mathbf{k}^2) \tag{10.83}$$

It will be convenient to divide the Y field into positive and negative frequency parts:

$$Y_u{}^+{}_i(y) = \int d^{(D-1)}k\, N_0(k) \sum_{\lambda=1}^{D-2} \varepsilon_i(k, \lambda)\, a(k,\lambda)\, e^{-ik\cdot y} \tag{10.84}$$

and

$$Y_u{}^-{}_i(y) = \int d^{(D-1)}k\, N_0(k) \sum_{\lambda=1}^{D-2} \varepsilon_i(k, \lambda)\, a^\dagger(k,\lambda)\, e^{ik\cdot y} \tag{10.85}$$

For later use we note the commutator between the positive and negative frequency parts is:

$$[\, Y_u{}^-{}_j(y_1), Y_u{}^+{}_k(y_2)] = -\int d^{(D-1)}k\, e^{ik\cdot(y_1 - y_2)}\, (\delta_{jk} - k_j k_k/\mathbf{k}^2)/[(2\pi)^{D-1}\, 2\omega_k] \tag{10.86}$$

10.3.3.1 Y^μ Fock Space Imaginary Coordinate States

States can also be defines for the quantized Y^μ field. These states will be similar in form to electromagnetic photon states but play a different role in our approach since they are in fact coordinate excitation states for the imaginary part of $Y^i(y)$ (eq. 10.4). Thus universe particles (and other fields) will exist in a real D-dimensional space with quantum excitations into imaginary Quantum Dimensions. These excitations become significant at high energies. At low energies space appears as c-number complex; at very high energies space becomes slightly q-number complex.

There are two types of imaginary coordinate excitations: 1.) Quantum excitations into Fock states consisting of a superposition of states with a definite finite number of Y_u "particles" and 2.) Imaginary coordinate excitations into coherent Y_u states with an "infinite" number of particles. Coherent states can be viewed as representing "classical" fields.

In this section we will consider Y_u field states with a definite number of excitations ("particles"). The raising and lowering operators of the Y_u field can be used to define free particle states. For example a one particle state can be defined by

$$|k, \lambda\rangle = a^\dagger(k, \lambda)|0\rangle \tag{10.87}$$

with corresponding bra state

$$\langle k, \lambda| = \langle 0|a(k, \lambda) \tag{10.88}$$

where the "coordinate vacuum" is defined as usual:

$$a(k, \lambda)|0\rangle = 0 \tag{10.89}$$

$$\langle 0|a^\dagger(k, \lambda) = 0 \tag{10.90}$$

Multi-particle states can also be defined in the conventional way with products of the raising and lowering operators applied to the vacuum. The set of all states containing a finite number of "particles" constitutes a Fock space.

A state with a finite number of Y_u "particles" represents a quantum fluctuation into imaginary Quantum Dimensions.

10.3.3.2 Y_u Coherent Imaginary Coordinate States

Coherent Y_u states bring us closer what we might consider to be "classical" imaginary dimensions – dimensions that we can, in principle, experience as we do normal dimensions. Let us define the coherent state[99]

$$| y, p\rangle = e^{-\mathbf{p}\cdot \mathbf{Yu}^-(y)/M_u^{D/2}}|0\rangle \tag{10.91}$$

This state is an eigenstate of the coordinate operator $Y_u^+(y')$:

$$Y_{u\ j}^{+}(y_1)\ |y_2, p\rangle = -[Y_{u\ j}^{+}(y_1), \mathbf{p}\cdot\mathbf{Y}^-(y_2)]/M_u^{D/2}|y, p\rangle \tag{10.92}$$

$$= -\int d^{D-1}k\ [N_0(k)]^2\ e^{ik\cdot(y_2 - y_1)}\ (p_j - k_j\mathbf{p}\cdot\mathbf{k}/\mathbf{k}^2)/M_u^{D/2}|y, p\rangle$$

---

[99] Coherent states are well known in the physics literature. See for example T. W. B. Kibble, J. Math. Phys. **9**, 315 (1968) and references therein; V. Chung, Phys. Rev. **140**, B1110 (1965); J. R. Klauder, J. McKenna, and E. J. Woods, J. Math. Phys. **7**, 822 (1966) and references therein.

$$= p^i\Delta_{Tij}(y_1 - y_2)/M_u^{D/2}|y, p> \tag{10.93}$$

where $p^i\Delta_{Tij}(y_1 - y_2)/M_u^{D/2}$ is the eigenvalue of $Y_u{}^+{}_j(y_1)$. As we will see later, the eigenvalue of $Y_u{}^+$ becomes large as $(y_1 - y_2)^2 \to 0$. Thus the imaginary Quantum Dimensions become significant at very short distances, and then significantly modifies the high-energy behavior of quantum field theories. In particular, Quantum Dimensions have a significant effect when

$$(y_1 - y_2)^2 \lessapprox (2^{D-2}\pi^{D-2}M_u{}^2)^{-1} \tag{10.94}$$

We assume the mass scale $M_u = M_C$ is very large – perhaps of the order of the Planck mass (1.221×10^{19} GeV/c^2).

10.3.3.3 Quantization of the Type I Free Universe Particle Dirac Field

The quantization procedure is formally identical to that of a conventional Dirac particle. The standard equal time anti-commutation relations for a D-dimensional fermion field are:

$$\{\psi_\alpha(Y), \psi_\beta(Y')\} = \{\pi_{\psi\alpha}(Y), \pi_{\psi\beta}(Y')\} = 0 \tag{10.95}$$

$$\{\pi_{\psi\alpha}(Y), \psi_\beta(Y')\} = i\,\delta_{\alpha\beta}\,\delta^{D-1}(\mathbf{Y} - \mathbf{Y'}) \tag{10.96}$$

where α and β are the spinor indices ranging from 1 to $N_{MRC} = 2^{D/2}$ and where

$$\pi_{\psi\alpha}(Y) = i\,\psi_\alpha{}^\dagger(Y) \tag{10.97}$$

The field can be expanded in a fourier series:

$$\psi(Y(y)) = \sum_s \int d^{D-1}p\; N^d{}_m(p)\, [b(p,s)u(p,s) :e^{-ip\cdot(y + iYu/M_u{}^{D/2})}: + d^\dagger(p,s)v(p,s) :e^{ip\cdot(y + iYu/M_u{}^{D/2})}:] \tag{10.98}$$

$$\psi^\dagger(Y(y)) = \sum_s \int d^{D-1}p\; N^d{}_m(p)\, [b^\dagger(p,s)\bar{u}(p,s)\gamma^0 :e^{+ip\cdot(y + iYu/M_u{}^{D/2})}: + d(p,s)\bar{v}(p,s)\gamma^0 :e^{-ip\cdot(y + iYu/M_u{}^{D/2})}:] \tag{10.99}$$

where

$$N^d{}_m(p) = [m/((2\pi)^{D-1}E_p)]^{1/2} \tag{10.100}$$

and

$$E_p = p^D = (\mathbf{p}^2 + m^2)^{1/2} \tag{10.101}$$

with : … : signifying normal ordering. The commutation relations of the Fourier coefficient operators are:

$$\{b(p,s), b^\dagger(p',s')\} = \delta_{ss'}\delta^{D-1}(\mathbf{p} - \mathbf{p'}) \tag{10.102}$$

$$\{d(p,s), d^\dagger(p',s')\} = \delta_{ss'}\delta^{D-1}(\mathbf{p} - \mathbf{p}') \tag{10.103}$$
$$\{b(p,s), b(p',s')\} = \{d(p,s), d(p',s')\} = 0 \tag{10.104}$$
$$\{b^\dagger(p,s), b^\dagger(p',s')\} = \{d^\dagger(p,s), d^\dagger(p',s')\} = 0 \tag{10.105}$$
$$\{b(p,s), d^\dagger(p',s')\} = \{d(p,s), b^\dagger(p',s')\} = 0 \tag{10.106}$$
$$\{b^\dagger(p,s), d^\dagger(p',s')\} = \{d(p,s), b(p',s')\} = 0 \tag{10.107}$$

The spinors u(p,s) and v(p,s) are defined in a conventional way (as in Bjorken and Drell). However their form is different from the 4-dimensional case. If one takes the $N_{MRC}\times N_{MRC} \equiv 2^{D/2}\times 2^{D/2}$ $\gamma\cdot p$ matrix, then the first $2^{D/2-1}$ columns give u(p,s) up to a normalization for the free particle case, the remaining $2^{D/2-1}$ columns give v(p,s) up to a normalization.

Since there are $2^{D/2-1}$ possible spin values, using the equation 2s + 1 = total number of spin values, we see that the spin of a fermionic universe particle is

$$s_M = 2^{D/2-2} - ½ = 2{,}147{,}483{,}647/2 \quad \approx 1 \text{ billion.}$$

The possible universe particle spin values are:

Up spin values: $+1/2^{D/2-1}, +2/2^{D/2-1}, \ldots, +2^{D/2-2}/2^{D/2-1}$

Down spin values: $-2^{D/2-2}/2^{D/2-1} = -½, -2^{D/2-2}/2^{D/2-1} + 1, \ldots, -1/2^{D/2-1}$

and amount to a total of about 2.5 billion spins.

This enormous number of possible spins is reasonable considering the number of dimensions D and the enormous variety in spins one should expect in universes – given their large size and complexity.

10.3.3.4 Feynman Propagators for the Type I Free Universe Particle Dirac Field

The form of the fermionic universe particle Feynman propagator differs from a conventional fermion propagator by having a Gaussian factor R(**p**, z) in its fourier expansions. This follows from using quantum Megaverse coordinates (eq. 10.4).

$$iS_F^{TT}(y_1 - y_2) = \langle 0|T(\bar{\psi}(Y(y_1))\psi(Y(y_2)))|0\rangle \tag{10.108}$$

where the time ordering is with respect to y_1^D and y_2^D. Expanding the free fields leads to the fourier representation:

$$iS_F^{TT}(y_1 - y_2) = i \int \frac{d^D p\, e^{-ip\cdot(y_1 - y_2)}\,(p + m)\, R(\mathbf{p}, y_1 - y_2)}{(2\pi)^D (p^2 - m^2 + i\varepsilon)} \tag{10.109}$$

where

$$R(\mathbf{p}, y_1 - y_2) = \exp[-p^i p^j \Delta_{Tij}(y_1 - y_2)/M_u^D] \tag{10.110}$$

$$= \exp\{-p^2[A(v) + B(v)\cos^2\theta] / [(2\pi)^{D-2} M_c^4 z^2]\} \tag{10.111}$$

(Note p^2 is the square of the spatial (D – 1)-vector.) with

$$z^\mu = y_1^\mu - y_2^\mu \tag{10.112}$$

$$z = |\mathbf{z}| = |\mathbf{y_1} - \mathbf{y_2}| \tag{10.113}$$

$$p = |\mathbf{p}| \tag{10.114}$$

$$v = |z^0|/z \tag{10.115}$$

$$A(v) = (1 - v^2)^{-1} + .5v \ln[(v - 1)/(v + 1)] \tag{10.116}$$

$$B(v) = v^2(1 - v^2)^{-1} - 1.5v \ln[(v - 1)/(v + 1)] \tag{10.117}$$

$$\mathbf{p}\cdot\mathbf{z} = pz\cos\theta \tag{10.118}$$

and $|\mathbf{p}|$ denoting the length of a spatial (D – 1)-vector $\mathbf{p}$ while $|z^0|$ is the absolute value of $z^0 \equiv z^D$.

As eq. 10.109 indicates, the Gaussian damping factor[100] R(p, z) for large spatial momentum p is the same for both the positive and negative frequency parts of the Two Tier Feynman propagator. We are assuming the spatial momentum is real-valued in this discussion. It is also important to note that R(p, z) does not depend on $p^0 = p^D$ (in the Y Coulomb gauge) and thus the integration over p^0 proceeds in the usual way to produce time-ordered positive and negative frequency parts.

10.3.3.5 Feynman Propagators for the Types II, III, and IV Free Universe Particle Dirac Fields

These propagators differ in details from the Type I propagator. The differences modulo the change in dimension appear in Blaha (2011c). See also Blaha (2005a) for a detailed discussion of 4-dimensional spin ½ particle propagators.

10.3.4 Expanding and Contracting Universes: Impact of Time Dependent Universe Particle Masses

Our discussions of the dynamics of universe particles assumed their masses were constant. However the definition of mass in terms of the area of a universe based on the physics of black holes is

[100] Note the Gaussian damping is for all D – 1 spatial momentum integrations.

$$M = \kappa A/8\pi \qquad (10.119)$$

where A is the area of the black hole shows that *the mass of a universe particle is time dependent* because the area of a universe is generally time dependent. For example, our universe is expanding and its surface area is thus growing with time.

Eqs. 10.11 (and subsequent fermionic dynamic equations) must then be modified from

$$\psi(y) = \exp[-imt]w(0) \qquad (10.11)$$

to a covariant form:

$$\psi(y) = \exp[-i\int_0^{w\cdot y} m(t')dt']w(0) \qquad (10.120)$$

where w is a unit D-vector in the time (y^D) direction ($w^2 = 1$). The lower bound on the integral, 0, is the time of the beginning of the universe particle – its Big Bang. Thus the cumulative change in the mass of the universe particle may be significant. It is interesting to note that the Wheeler-Dewitt equation also has a variable value mass term R that also depends on the evolution of the universe.

Eq. 10.120 satisfies the free covariant Dirac-like universe particle field dynamic equation

$$[i\gamma^i\partial/\partial y^i - m(w\cdot y)]\psi(y) = 0 \qquad (10.121)$$

In contrast to the constant mass equation eq. 10.19. Substituting eq. 10.120 in eq. 10.121 we find

$$(\gamma^i w_i\, m(w\cdot y) - m(w\cdot y))\psi(y) = 0 \qquad (10.122)$$

or

$$(\gamma^i w_i \;- 1)\psi(y) = 0 \qquad (10.123)$$

Upon performing a D-dimensional Lorentz boost (of the type of eqs. 10.13 – 10.16) on eq. 10.123 we obtain

$$(\gamma_i p^i/m_0 - 1)\psi(y) = 0$$

or

$$(\gamma_i p^i - m_0)\psi(y) = 0 \qquad (10.124)$$

where p^i is a momentum D-vector with $p^2 = m_0^2$. Eq. 10.123 is the constant mass momentum space dynamic equation. It determines the spinor in $\psi(y)$. After taking account of the quantum coordinates the quantum Dirac-like universe particle wave function has the form

$$\psi(Y(y)) = \sum_s \int d^{(D-1)}p\, N^d{}_m(p)\, [b(p,s)u(p,s) : \exp[-iG(p, Y(y))]: + d^\dagger(p,s)v(p,s) \cdot$$
$$\cdot :\exp[+iG(p, Y(y))]:\} \qquad (10.125)$$

$$\psi^\dagger(Y(y)) = \sum_s \int d^{(D-1)}p\, N^d{}_m(p)\, \{b^\dagger(p,s)\bar{u}(p,s)\gamma^0 :\exp[+iG(p, Y(y))]: + d(p,s)\bar{v}(p,s)\gamma^0 \cdot$$
$$\cdot \exp[-iG(p, Y(y))]:\} \qquad (10.126)$$

where : … : denotes normal ordering and

$$G(p, Y(y)) = \int_0^{p\cdot Y(y)/\lambda} m(t')dt' \qquad (10.127)$$

with $\lambda = m_0$, and $N^d{}_m(p)$ a normalization constant. Contrast eqs. 10.125-10.126 to the constant mass case eqs. 10.98-10.101. The *constant mass case* simply sets $m(t') = m_0$.

If we examine the integral eq. 10.127 for a short time interval δt in the particle's rest frame then $G(p, Y(y)) \approx m(0)\delta t$ and so we define $m(0) = m_0$. Based on the formula for universe particle mass (eq. 10.119) we anticipate that m_0 might be as large as the Planck mass or larger – thus an extremely short radius. Blaha (2013) describes a quantum Big Bang model in which the initial radius of the universe is $O(EM_{Planck}{}^{-2})$ where E is of the order of 1 and has the dimensions of [mass].

Thus we have a closed form definition of a quantum universe particle wave function for universe particles of type I. A similar procedure can be followed for universe particles of types II, III, and IV.

The Feynman propagator for type I quantum fields is *not* eq. 10.109 but now has a form reflecting the Y(y) dependence of the quantum fields in eqs. 10.125 and 10.126:

$$iS_F{}^{TT}(y_1, y_2) = i \int \frac{d^Dp\ \{ <0|\theta(y_{1D} - y_{2D})G(y_1, y_2) + \theta(y_{2D} - y_{1D})G(y_2, y_1)\}0>}{(2\pi)^D\, (p - m_0)} \qquad (10.128)$$

where p^D is the energy and

$$G(y_1, y_2) = : \exp[-iG(p, Y(y_1))]: :\exp[+iG(p, Y(y_2))]: \qquad (10.129)$$

Let

$$G_{tot}(y_1, y_2) = <0|\theta(y_{1D} - y_{2D})G(y_1, y_2) + \theta(y_{2D} - y_{1D})G(y_2, y_1)\}0> \qquad (10.130)$$
$$= <0|\theta(y_{1D} - y_{2D}):\exp[-iG(p, Y(y_1))]::\exp[+iG(p, Y(y_2))]: +$$
$$+ \theta(y_{2D} - y_{1D}) :\exp[-iG(p, Y(y_2))]::\exp[+iG(p, Y(y_1)):]|0>$$

$$= \langle 0|\theta(y^D{}_1 - y^D{}_2)\colon \exp[-i\int_0^{p\cdot Y(y1)/\lambda} m(t')dt']\colon\colon \exp[+i\int_0^{p\cdot Y(y2)/\lambda} m(t')dt']\colon +$$

$$+\ \theta(y^D{}_2 - y^D{}_1)\colon \exp[-i\ \exp[+i\int_0^{p\cdot Y(y2)/\lambda} m(t')dt']\colon\colon \exp[+i\int_0^{p\cdot Y(y1)/\lambda} m(t')dt']\colon|0\rangle$$

with $\lambda = m_0$ then

$$iS_F{}^{TT}(y_1, y_2) = i\int \frac{d^Dp\ G_{tot}(y_1, y_2)}{(2\pi)^D\ (\not{p} - m_0)} \qquad (10.131)$$

Except for the case of a constant mass, where $m(t) = m_0$, the Feynman propagator is not a function of $y_1 - y_2$. The evaluation of eq. 10.130 in the general case of a variable mass is straightforward but cumbersome. For the special case of a linear time dependence of the mass, $m(t) = at$, we find eq. 10.130 gives

$$G_{tot}(y_1, y_2) = \langle 0|\theta(y^D{}_1 - y^D{}_2)\colon \exp[-ia(p\cdot Y(y_1)/m_0)^2/2]\colon\colon \exp[+ia(p\cdot Y(y_2)/m_0)^2/2]\colon +$$
$$+\ \theta(y^D{}_1 - y^D{}_2)\colon \exp[-ia(p\cdot Y(y_2)/m_0)^2/2]\colon\colon \exp[+ia(p\cdot Y(y_1)/m_0)^2/2]\colon|0\rangle \qquad (10.132)$$

yielding a complex function of p, y_1, and y_2. *Note that the lower bound of the integrals in the Feynman propagator cancel and thus the need for an understanding of the beginning of a universe is removed in this case.*

We have shown that universe particle theory can handle the case of a variable universe mass m(t). Expanding or contracting (or oscillating) universe particles correspond to expanding and contracting (or oscillating) universes.

10.3.5 Left-Handed and Right-handed Universe Particles

In sections 10.3.1 and 10.3.2 we found that left-handed and right-handed tachyonic universe particles existed. The tachyonic nature of the universe particles indicates that their speed in the universe exceeds the "speed of light" of the Megaverse. The physical meaning of the handedness of these types of universes is an interesting issue. When we consider our universe we see left-handedness in the weak interactions of elementary particles. In addition it appears that organic molecules overwhelmingly favor left-handedness on earth although right-handed molecules exist in outer space and can be created in the laboratory. Right-handed molecules transform into left-handed molecules in watery media through electromagnetic effects.

Why nature favors left-handedness is an open question. It has given rise to speculations that gravitation, especially quantum gravitons, may be left-handed. The European Space Agency's Planck telescope will study polarization effects in the cosmos and may well be able to show that the gravitons starting from the beginning of the universe, and magnified by inflation in the universe's expansion, may be left-handed.

If handedness of gravitation is verified experimentally, then our theory of left-handed/right-handed universe particles would be supported. *Our universe would then be tachyonic and probably left-handed. We, in the universe, would, of course, not know of the velocity of the universe in the Megaverse.*

10.3.6 Internal Structure of Universe Particles

We have treated universes as particles in the preceding discussion taking an extremely large view of Megaverse particles just as elementary particle theory viewed nucleons at low energies (large distances). Now we develop a detailed view of universe particles in a manner analogous to the high energy view of the internal dynamics of nucleons that led to the quark-parton model of nucleons. In the present case we shall see that high energy Baryonic and other field probes of universe particles can yield a model of the internal structure of universes.

We know that universes are composed of matter and radiation. We believe that there is at least one possible accessible interaction between universes dependent on baryon number – a baryonic, D-dimensional gauge field. There are also other particle number gauge fields – but these are less likely to be significant since Dark matter has yet to be found except through its gravitational effects. In this section we will discuss the use of a baryonic gauge field to probe the baryon structure of universe particles.

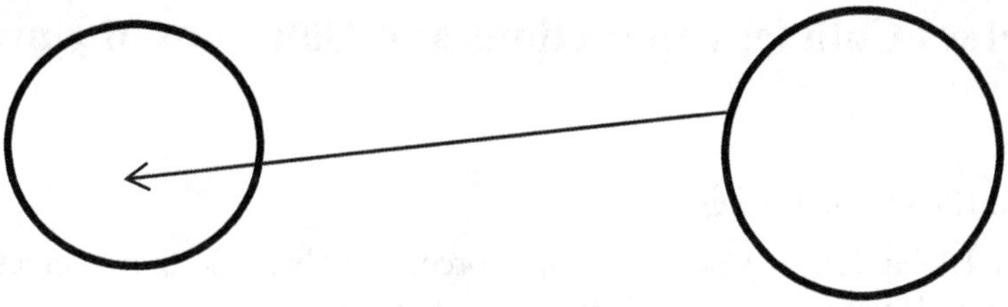

Figure 10.1. A symbolic view of a high resolution (high energy) probe from a universe to a specific baryonic part of another universe.

Figure 10.2. A symbolic view of a low resolution (lower energy) probe from a universe to an entire universe.

There appears to be two types of probes[101] of a universe: 1) a series of high energy probes to specific small regions inside another universe for the purpose of mapping its internal structure; 2) a low energy probe of another universe to get a global view of its structure. The first type of probe corresponds to deep inelastic (high energy) electron-nucleon scattering which led to the quark-parton model of nucleons. The second type of probe corresponds to low energy electron-nucleon scattering to get a "global" view of a nucleon. In both case an electromagnetic (gauge) field particle (photon) was the probe particle.

Besides the inherent scientific interest in such experiments it is possible that they may be of use in the very distant future if Mankind is able to develop Megaverse starships that can travel in the Megaverse to other universes. Then the baryonic gauge field may become the "eyes" of the starship in addition to the electromagnetic field (light) are the eyes of current spaceships. We considered the possibility of universe starships in the book entitled, *All the Megaverse! II* in detail, and in a later chapter of this book.

10.4 When Universes Collide: Interactions and Collisions of Universe Particles

10.4.1 Gravitation and the Other Forces

As we saw in Blaha (2015a) some of the forces involved in the interactions and collisions of universe particles are the force of gravity, and an additional set of forces which we take to be the listed in chapter 1.

We describe the baryonic force as an example in sections 10.11-10.15 below. The other forces are similar so we will not describe them in any detail here (See chapter 8 for more detail.) Massless gauge field interactions are of particular interest since they are long range. For example, the two baryonic forces are long range forces, both inside universes and in the Megaverse, because their gauge fields are massless.

[101] It would appear that the probes are in Megaverse coordinates, not universe coordinates, in order to 'bridge' the distance between universes. Chapter 8 shows the manner of the mapping between Megaverse coordinates and universe coordinates for quantum fields.

10.4.2 Universes in Collision

We assume that the dynamics of universes in collision will be analogous to that of galaxies in collision since gravity is a dominant force in both cases. Colliding galaxies have often been observed. Their dynamics should provide guidance for the case of universes in collision.[102]

It is clear in the case of colliding galaxies, and of colliding large nuclei (gold and lead typically) that there are several types of collisions with differing results. These types of universe collisions can be qualitatively classified as

1. Clean collisions in which universes nudge each other but retain their identity. These are extreme peripheral collisions. If the universes overlap slightly then the typically spherical symmetry of the universes may become distorted and they may become lopsided.[103]

2. Peripheral collisions in which the universes retain their identity but are connected by a trailing string of mass-energy. Eventually the string breaks and the universes separate. Subsequently the pieces of trailing string in each universe contract due to their universe's gravitational effects.

3. Two universes can collide and produce multiple universes.

4. Two universes can collide in a "central" collision and amalgamate into one universe.

We will discuss universe interactions from this viewpoint in more detail later.

10.5 Bosonic Universe Particles

The previous sections has described fermionic universe particles. In this section we will briefly describe aspects of bosonic (spin 0) universe particles. First it is important to note that the Wheeler-DeWitt equation being second order like the Klein-Gordon equation seems to suggest that universe particles can be bosonic – like Klein-Gordon equation particles.[104] The Wheeler-DeWitt equation has a mass-like term R that can be positive or negative. If the mass

---

[102] The high energy collision of atomic nuclei at Brookhaven, CERN and other laboratories also is analogous in overall detail with universes in collision.

[103] The Wilkinson Microwave Anisotropy Probe (WMAP) and the Planck European Space Agency satellite has been accumulating data since 2001 that suggests the universe may be lopsided with hot and cold spots on opposite sides of the universe differing from those on the other side being hotter and colder respectively. *Perhaps the result of a collision when the universe was young.*

[104] One should remember that the Wheeler-DeWitt equation is not in space but in a 6-dimensional manifold, denoted M, of metrics with one "time" dimension – having hyperbolic signature – + + + + + when the metric is positive definite. See DeWitt's paper.

term is negative then the wave-like propagation of the state functional (wave equation solution) can be in space-like directions implying a tachyonic solution. Thus the Wheeler-DeWitt equation supports "normal" state functionals that propagate in time-like directions as well as tachyonic propagation.

For this reason we suggest that bosonic universe particles can be either normal or tachyonic. Tachyonic bosonic universe particles can fission in a manner similar to tachyonic fermionic universe particles. The fission equations of section 10.5.2 also apply to tachyonic bosonic universe particles.

The quantum field theory of normal and tachyonic bosonic universe particles is similar to that of ordinary bosons. See Blaha (2005a) for the boson case discussion that is paralleled by our universe particle formalism.

10.6 Physical Meaning of Universe Particle Spin

The physical meaning of spin is a continuing discussion topic. We have suggested[105] that spin states are in essence logic states with changes in spin an analogous to changes in logical values in a discourse or computer program. Since the matrix formalism for spin ½ and higher spin states is formally similar to the formalism for angular momentum, one can combine spin and angular momentum as we do in quantum theory.

In the case of universe particles, one can also associate universe particles with "true" and "false" values. Fermionic universe states have $2^{D/2 - 1}$ truth values and correspond to a multi-valued logic. The numerousness of truth values is due to the D-dimensional space within which universe particles reside.

Naturally one would like to know the physical differences between these $2^{D/2 - 1}$ types of universe particles. Does the difference reside in different shapes of the universe particles? Or is the difference somehow a consequence of the global mass-energy distribution of the universe that we have not been able to discern since we only know of one universe?

The physical meaning of spin for elementary particles is also somewhat elusive. It does not reflect the flow of charge within a particle. For if it did reflect physical spinning of a particle, the outer edges of a particle such as an electron would be traveling at a speed faster than light. So spin is not a mechanical property of the internal structure of an elementary particle. We have suggested that it is a truth value in the matrix formulation of a 4-valued logic called Asynchronous Logic. Thus it has no certain tangible physical basis.

In the case of universe particles the situation is unclear at present. It could be taken to be an indirect reflection of the structure of mass-energy within a universe. This view would be contrary to our proposed view of elementary particle spin as truth values. So we can only assert that a logic interpretation is the only sensible one (based on our present knowledge or our lack thereof). The physical role of universe particle spin is only evident in interactions between universe particles via gauge fields. Thus one must simply view it as a construct for the present.

---

[105] Blaha (2011c) and subsequent books.

Other than our mapping of spin values to logic values in Asynchronous Logic there is little anyone can say about the physical origin of elementary particle spin. Specifying a symmetry group as the origin is not sufficient.

10.7 Elementary Particles with Time Dependent Masses

The discussions of this chapter were presented for universe particles. However they could also apply to elementary particles or condensed matter excitations with some changes – primarily changes due to different dimensions.

Elementary particles with time dependent masses do not exist as far as we know. And that is a good thing. The idea of an elementary particle expanding indefinitely, like a universe, would have disastrous consequences for life, as would particles contracting indefinitely. However, particles with oscillating masses would seem to be physically acceptable. The development of an elementary theory with oscillating masses would be an interesting exercise that has applications in condensed matter physics.

10.8 Impact of Universe Particle Acceleration – Lopsided Internal Structure of Universe

We are developing a theory of universe particle interactions. Such interactions would cause universe particles to accelerate and should be detectable within a universe as a "lopsidedness" – there would be a shift of parts of the universe away from the direction of acceleration resulting in a difference in the features of the universe "in front" compared to those "in back" – an acceleration effect just as one sees when a jet accelerates.

Interestingly new data from the Planck observatory of the European Space Agency confirms and extends earlier data from NASA's WMAP observatory that one side of the universe appears different from the other side. There are temperature differences and mass distribution differences – just as one might expect if the universe were accelerating as a unit.

Thus we see the beginning of data suggesting our universe may be moving – "indeed accelerating" – through a Megaverse. Some Planck observatory scientists have suggested their data is a preliminary indication of the Megaverse.

10.9 Megaverse Baryonic Gauge Field - Plancktons

The conservation of baryon number has been repeatedly investigated by experimenters and found to be true to extremely high accuracy. For decades theorists have suggested that the conservation law follows from the existence of a gauge field in a manner much like electric charge conservation follows from the properties of the electromagnetic abelian gauge field.[106]

We will therefore assume a baryonic gauge field exists that is similar to the electromagnetic field except for features due to its definition and existence in the D-dimensional Megaverse. This field will couple extremely weakly to individual baryons as well as universe

---

[106] See Gell-Mann, M. and Levy, M. *Nuovo Cimento* 16, 705 (1960) for a proof.

particles with non-zero baryon number. We will call the baryonic gauge field particle a *planckton*. Its electromagnetic analogue is the photon. We described D-dimensional baryonic gauge field quantization in Blaha (2015a) and earlier books.

Plancktons propagate in the Megaverse, both within universes, and in the Megaverse external to universes. So we will define the planckton field in D-dimensional Megaverse coordinates. They will interact with baryons within a universe with Megaverse coordinates mapped to the curvilinear coordinates in the universe. (This mapping was discussed earlier.)

Since a planckton field in D-dimensional conventional coordinates would lead to divergences we will use quantum coordinates:

$$Y^i(y) = y^i + i\, Y_u{}^i(y)/M_u{}^{D/2}$$

with quantum coordinate derivatives defined by

$$\partial_i = \partial/\partial Y^i(y) = \partial/\partial(y^i - Y_u{}^i(y)/M_u{}^{D/2})$$

to obtain a completely finite theory of planckton interactions with elementary particles and universe particles.

Plancktons, particle fields, universe fields, gravitation, other gauge fields, and the $Y_u{}^i(y)$ field of quantum coordinates are the only fields in the space between universes in the Megaverse. Universe Extended Standard Model gauge fields are confined to a universe. They are thus zero in the Megaverse space between universes.

However, we believe all gauge fields in the Extended Standard Model have a matching analogue in the Megaverse. These Megaverse fields do not have any continuity conditions relating them to their counterparts in universes.Universe fields are confined to universes. Their Megaverse equivalents are spread throughout the Megaverse including inside universes. Megaverse fields must be very small to have avoided detection in astronomical and earth-based experimental studies.See chapter 8 for a more detailed discussion.

10.10 Beyond the Planckton

The analogy between plancktons and electromagnetic field photons raises the possibility that the baryonic gauge field may be a gauge field. We showed that the four particle number interactions embody a broken U(4) symmetry within universes in I. Their U(4) symmetry gives rise to the four fermion generations.

10.11 Planckton Interactions with Universe Particles and Individual Baryons

Section 15,1 of Blaha (2015a) describes the second quantization of plancktons in 16 dimensions. In this section we will develop an interacting theory in D dimensions of universe particles and plancktons from the model lagrangian terms of universe particles, plancktons and quantum coordinates. We will only consider the case of type I universe particles since the other cases differ from it only in details. The universe particle – planckton lagrangian terms are:

$$\mathcal{L} = \bar{\psi}(Y(y))[i\gamma^{\mu}\partial/\partial y^{\mu} - e_B\gamma^{\mu}B_{u\mu}(Y(y)) - m(t)]\psi(Y(y)) - \tfrac{1}{4} F_{Bu}{}^{\mu\nu}(Y(y))F_{Bu\mu\nu}(Y(y)) - \\ - \tfrac{1}{4} F_u{}^{\mu\nu}(y)F_{u\mu\nu}(y) \tag{10.133}$$

where $\mu, \nu = 1, 2, \ldots, D$ and where

$$\bar{\psi} = \psi^{\dagger}\gamma^{D}$$

$$F_{Bu\mu\nu} = \partial B_{u\mu}(Y(y))/\partial Y^{\nu}(y) - \partial B_{u\nu}(Y(y))/\partial Y^{\mu}(y) \tag{10.134}$$

$$F_{u\mu\nu} = \partial Y_{\mu}/\partial y^{\nu} - \partial Y_{\nu}/\partial y^{\mu}$$

$$Y^{i}(y) = y^{i} + i\, Y_u{}^{i}(y)/M_u{}^{D/2}$$

$$e_B = e_{B0}/M_u{}^{D/2-2}$$

with e_{B0} a dimensionless coupling constant, and with μ and ν ranging from 1 through D.

The corresponding lagrangian is

$$L = \int d^{(D-1)}y\,\mathcal{L} \tag{10.135}$$

Note the dimensions of the fields differ in the D dimensional space:

$$Y^{\mu} \sim [\text{mass}]^{D/2-1}$$

$$B_{u\mu} \sim [\text{mass}]^{D/2-1}$$

$$\psi \sim [\text{mass}]^{(D-1)/2}$$

as can be seen from the above lagrangian as well as earlier equations. Note also that the mass and thus the size of universe particles is time dependent. They can expand or contract with time depending on their internal characteristics (gravitation and effects of elementary particle interactions) which are not embodied in this lagrangian. As a result, this theory, incomplete as it is, does not conserve energy unless m(t) is constant.

The lagrangian generates the baryonic interactions of universe particles using Two Tier quantum coordinates which prevent infinities in perturbation theory calculations.

The interaction of baryon elementary particles with the baryonic field requires terms in Extended Standard Model covariant derivatives specifying the baryon field interaction baryons with the form

$$e_B\gamma^{\mu}B_{u\mu}(Y(y))$$

The following sections describe some of the physically significant interactions that the lagrangian implies.

10.12 Creation of Universes through Gauge Field Fluctuations

One of the most exciting questions in Cosmology is the origin of our universe. The conventional view is that it originated in a Big Bang from an infinitesimal point in space. The source of the Big Bang and the prior state of the Cosmos, if there was one, is the subject of much speculation. Based on the particle interpretation of the Wheeler-DeWitt equation, the possibility of a baryonic force strongly supported by conservation of baryon number, and the Megaverse concept, it is reasonable to consider the possibility that the universe originated in a vacuum fluctuation.

In this case there would be two Big Bangs one for our universe and one for an anti-universe. One would expect that they would have reverse corresponding features: one with baryon dominance – one with anti-baryon dominance, and one left-handed – one right-handed.

Our formulation of universe particle theory provides for the generation of a universe particle and anti-particle as a vacuum fluctuation. We view a universe particle as having a substantial excess of baryons, N, as we see in our universe. Its anti-universe at the time of creation (the Big Bang point) is its "mirror image" having the "same" number of anti-baryons (baryon number –N) so that baryon number is conserved by the fluctuation event. Thus the excesses of one is compensated by the excesses of the other.

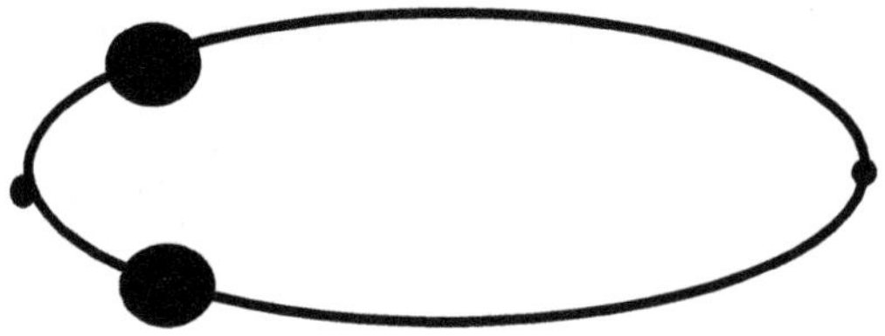

Figure 10.3. Generation of a universe – anti-universe pair as a vacuum fluctuation.

The small value of the coupling constant should lead to an extremely long lifetime for the universes generated by the fluctuation. Thus the 13.7 billion year life of our universe is not unreasonable. The probability of the creation of universes by vacuum fluctuations should be correspondingly small.

10.13 When Universes Collide: Coalescence of Universes

Universes moving in the Megaverse can collide through chance, or due to the planckton field which can cause universes with excess baryons to attract universes with excess anti-baryons.

When universes collide several possibilities present themselves:

1. They can graze each other distorting each other's shape and internal baryon distribution through the baryonic and other forces while maintain their individual identity.

2. They can intermix with both the baryonic, gauge, and gravitational forces causing a redistribution of their masses. They may separate afterwards or may coalesce into a single universe. One result of this may be lopsided universes. Our universe appears to be lopsided. Some cosmologists believe this is due to a near collision of our universe with another shortly after the Big Bang.

In our discussion we have been referring only to the planckton baryon field for the sake of concreteness. But the three other massless particle number fields will also play a role in universe interactions. The relative strength of these interactions is not known since we do not know the size of their coupling constants.

10.14 Fission of Universes

Under certain circumstances the distribution of matter in the universe may lead to the fission of the universe into two separate universes. Our model lagrangian supports this possibility for universe particles. The detailed mechanism of the fission process is not specified by the model.

10.14.1 Fission of Normal universes

The fission of universe particles in our universe particle model is depicted in the Feynman diagram in Fig. 10.4.

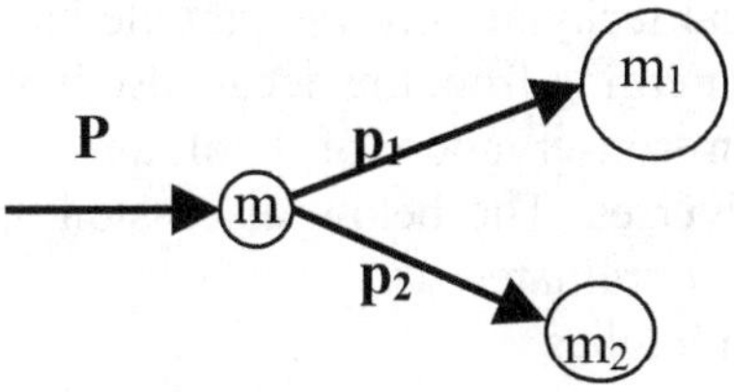

Figure 10.4. Fission of a universe particle into two universe particles.

The sum of the masses of the output universe particles is usually less than the original universe particle mass. However if the fission takes a long time and the masses are time dependent then the produced universe particles combined masses may exceed the original universe's mass.

10.14.2 Tachyon Universe Particle Fission to More Massive Universe Particles

In Blaha (2007a) we showed that a tachyonic (faster than light) particle could fission into particles of larger mass through the conversion of momentum into mass. In this section we will show that a tachyonic universe particle may fission into two more massive universe particles.[107] This phenomenon is of particular interest because it enables tachyonic universes to spawn in a new novel way not previously considered in discussions of the origin of universes.

A model lagrangian for a tachyonic universe particle is

$$\mathcal{L}_{\parallel} = \psi_T^S(Y(y))[\gamma^\mu\partial/\partial y^\mu - e_B\gamma^\mu B_{u\mu}(Y(y)) - m(t)]\psi(Y(y)) - \tfrac{1}{4} F_{Bu}^{\mu\nu}(Y(y))F_{Bu\mu\nu}(Y(y)) - \\ - \tfrac{1}{4} F_u^{\mu\nu}(y)F_{u\mu\nu}(y) \quad (10.136)$$

We assume m(t) is constant.

When a particle or a universe particle fissions (decays) one normally expects that the masses of the particles or universe particles produced by the decay to be smaller than the mass of the original particle or nucleus. In the case of tachyonic (faster-than-light) elementary particles or universe particles a much different possibility is present: a tachyon can decay into heavier tachyons. We will consider the specific case of a tachyon universe particle decaying into two universe particles whose total mass is greater than the original. (See Fig. 10.5.)

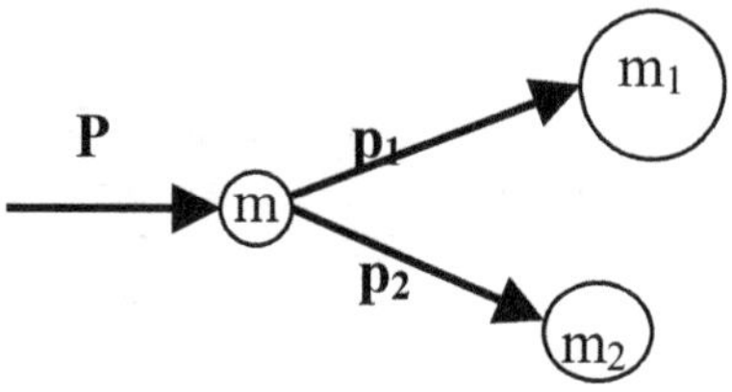

Figure 10.5. Two universe particle decay of a tachyon universe particle.

We will assume the initial tachyon universe particle has zero energy ($p^D = 0$) and thus the tachyons universe particles emerging from the decay also have total universe particle energy zero. The analysis is based on conservation of total universe energy and momentum in Megaverse space outside of universes. The below discussion applies to D-dimensional space with (D – 1)-dimensional spatial coordinates.

Momentum conservation implies

$$\mathbf{P} = \mathbf{p}_1 + \mathbf{p}_2 \quad (10.137)$$

Since all energies are zero

$$(cP)^2 = (c\mathbf{P})^2 = m^2$$
$$(cp_1)^2 = (c\mathbf{p}_1)^2 = m_1^2 \quad (10.138)$$
$$(cp_2)^2 = (c\mathbf{p}_2)^2 = m_2^2$$

107 We will use the term mass here to denote mass-energy. Since we identified mass as a multiple of area earlier the comments here would appear to apply to universe area as well.

where P = |**P**|, p_1 = |$\mathbf{p}_1$|, and p_2 = |$\mathbf{p}_2$|. If we now square eq. 10.137 and then use eqs. 10.138 we obtain

$$m^2 = m_1^2 + m_2^2 + 2m_1m_2 \cos\theta \qquad (10.139)$$

where θ is the opening angle between the emerging universe particles momenta $\mathbf{p}_1$ and $\mathbf{p}_2$. Eq. 10.139 has a number of interesting cases:

Case θ = 0:

$$m = m_1 + m_2 \qquad (10.140)$$

The masses of the outgoing universe particles sum to the mass of the original tachyon universe particle.

Case θ = π/2:

$$m^2 = m_1^2 + m_2^2 \qquad (10.141)$$

The masses of each outgoing universe particle tachyon is less than the mass of the original tachyon universe particle.

Case θ = π:

$$m^2 = (m_1 - m_2)^2 \qquad (10.142)$$

In this case either $m_1 > m$ or $m_2 > m$. Thus one of the outgoing tachyon universe particles has a greater mass than the original tachyon universe particle. Mass is effectively created from the spatial momentum of the initial universe particle. This process is the inverse of normal particle and universe particle fission where the sum of the outgoing masses is always less than the original particle's mass and the difference is mass converted into energy in the form of additional photons.

This last case, where one of the outgoing universe particles is more massive than the original universe particle, is not just for θ = π. Since

$$\cos\theta = (m^2 - m_1^2 - m_2^2)/(2m_1m_2) \qquad (10.143)$$

we see that the sum of the outgoing universe particle masses is always greater than the original tachyon universe particle *mass (except when* θ = *0)* since

$$\cos\theta = 1 + [m^2 - (m_1 + m_2)^2]/(2m_1m_2) \le 1 \qquad (10.144)$$

and thus

$$[m^2 - (m_1 + m_2)^2]/(2m_1m_2) \le 0 \qquad (10.145)$$

Note $m = m_1 + m_2$ only if $\theta = 0$.

Since we can transform the above discussion to the case of universe particle tachyons having non-zero Megaverse energy using an ordinary D-dimensional Lorentz transformation the discussion in this subsection is general.

We therefore conclude that when a tachyon universe particle decays into two tachyon universe particles the sum of the masses of the produced tachyon universe particles is greater than the mass of the original tachyon universe particle except if the angle between the momenta of the produced tachyon universe particles is zero. In that case the sum of the masses of the produced tachyon equals the mass of the original tachyon universe particle and the produced universe particles overlap.

10.15 Universe Particle – Planckton Interactions

These interactions are quite similar to Two Tier electromagnetic interactions except that universe particles have time-dependent masses, and that the space is D-dimensional.

The interactions have a new aspect due to the time dependence of the universe particle masses. This feature is illustrated by Fig. 10.6: the mass of a universe particle after a baryonic interaction vertex is the same as it was before the interaction assuming the point-like interaction specified in the lagrangian.

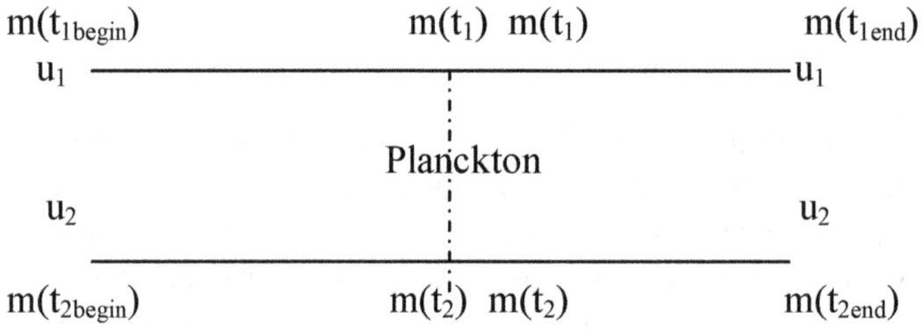

Figure 10.6. A Feynman diagram illustrating the continuity of a universe particle mass through a Planckton interaction.

The reader may verify this by writing the perturbation theory equivalent. A universe particle vertex corresponds to

$$iS_F^{TT}(y_1, y_2)\gamma^\mu iS_F^{TT}(y_2, y_3) \tag{10.146}$$

The universe particle mass is the same on either side of the interaction vertex.

10.16 Internal Structure of Universe Particles

We have developed a beginning in planckton field theory that gives interactions between baryons. This theory is applicable to universe-universe interactions. It also yields baryon particle – baryon particle interactions as well as baryon particle – universe particle interactions.

It is possible for a planckton to be emitted in one universe and interact with a baryon elementary particle in another universe. This type of "probe" must be a high energy probe just as a photon probe of the internal structure of a nucleon[108] must be a high energy photon to bring out the nucleon's internal structure (parton model).

In this section we will discuss planckton probes of other universes, and the internal structure of a universe as a mass distribution governed by gravitation as it relates to universe particles.

10.16.1 Planckton Probes

Plancktons can be generated in one universe and be used to probe the baryon distribution of another universe. Since the planckton propagator is expressed in Megaverse coordinates the baryon distribution in the target universe will be a distribution in Megaverse coordinates. Megaverse coordinates can be expressed in terms of the curved space-time coordinates of a universe x^μ. However the inversion of the map between universe coordinates and Megaverse coordinates

$$x^\mu = f^{-1\mu}(y) \tag{10.147}$$

is not 1:1 since x^μ is 4-dimensional and y is a D-dimensional vector. The universe coordinates x^μ are each individually determined up to a subspace. One might be concerned about this situation but the determination of the distribution in Megaverse coordinates gives a more direct picture not convoluted by the curvature of the target universe.

The detailed probing of a target universe requires high energy plancktons. The similarity of this procedure to deep inelastic electron-nucleon scattering is obvious to the high energy physicist. But in doing this planckton probe experiment one obtains a picture of a different universe – something that is not possible to do with electromagnetic or graviton probes.

10.16.2 Internal Structure of a Universe Particle

The development of the theory of universe particles which resulted in the lagrangian of eq. 10.136 does not fully describe universe particles since it neglects the internal structure of a universe particle. The internal structure of a universe particle is primarily determined by gravitation, electromagnetic effects and nuclear physics.

Consequently the full lagrangian of a universe particle has the form

$$\mathscr{L}_{tot} = \mathscr{L}_{internal} + \mathscr{L} \tag{10.148}$$

[108] Deep inelastic electron-nucleon scattering.

where $\mathcal{L}$ is determined above. As a result the complete quantum wave function of a universe particle has the form

$$\psi_{tot} = \psi_{internal}(Y)\psi_{ext}(Y) \tag{10.149}$$

where $\psi_{internal}(Y)$ is the internal wave function and $\psi_{ext}(Y)$ is external wave function. It seems reasonable to have a separable equation except when universes collide. In that situation a perturbative mixing of the universes and their wave functions applies and it may be possible to calculate the collision output universes by introducing a further interaction between the internal and external aspects of the universe particles.

10.17 Central Role of the Baryonic Force for Travel into the Megaverse

We have focused on the baryonic force in this chapter although the Dark baryonic force and the lepton forces may well play a role in universe particle dynamics. The reason is the long range possibility of travel into the Megaverse to other universes using the baryonic force to exit our universe into D-dimensional space using a slingshot orbit around a neutron star. We discuss this possibility later and refer the interested reader to Blaha (2014c) for a detailed discussion.

11. Dark Energy and an Inflationary Big Bang

In I we finished the derivation/construction of the form of the Extended Standard Model based primarily on the use of complex space-time coordinates and the Reality group. It includes Dark Matter and a justification for fermion generations based on a "new" broken symmetry, the Generation group symmetry. We introduced a universal field Y^μ that eliminated the divergences that appear in perturbation theory without the use of a renormalization program. We now turn to the extension of our theory to the Cosmology of our universe, and the extension of Nature to include other universes, which together with our universe, reside in a D dimensional Megaverse.

It seemed appropriate to study the evolution of our universe from its beginning, and view it as a model of the general evolution of universes in the Megaverse. There does appear to be any reason to view our universe as unique. Given the oft noted tendency of Nature to repeat the characteristics of phenomena we believe our universe should provisionally be viewed as a 'typical' universe.

Almost all of this chapter,[109] and 11A, 12, and 13 following, first appeared in Blaha (2004).

11.1 Dark Energy

Dark Energy surfaced in Cosmology when it was noticed that the speed of expansion of our universe was expanding due to an unkown, and "undetectable" energy, dubbed Dark Energy.A theoretical framework was developed by A. Guth and others called *inflation* that provided a scenario for the expansion of the universe through an unidentified source called Dark Energy.

We shall now show that the cause of inflaton is the $Y^\mu(y)$ quantum field of our Extended Standard Model that also resolved QFT divergence problems. We will see $Y^\mu(y)$ also makes the Big Bang finite—no singularity; as well as freeing The Standard Model and Quantum Gravity[110] of infinities. *Thus $Y^\mu(y)$ has a remarkable triple role in our view – to eliminate the Big Bang singularity, to generate the explosive growth of the universe, and to remove infinities from The Standard Model and Quantum Gravity.* This happy coincidence of solutions reflects the use of Ockam's Razor to find the simplest general solution and Leibniz's Minimax Principle

---

[109] While the calculation is unchanged we have added clarifications of the original that might have led to misunderstandings particularly in clarifying the constant a vs. the scale function a(t).

[110] See Blaha (2011c) and Blaha (2005a) for the removal of infinities in Quantum Gravity.

to find the most minimal solution that has maximal effects – two sides of the same coin in a sense.

Since the $Y^\mu(x)$ quantum gauge field is a free field (neglecting gravity) the initial state of the universe can be permeated with quanta of this field as well as particle quanta. The total energy of the free $Y^\mu(x)$ field within the universe is Dark Energy.

11.2 The Big Bang Experimentally

In the light of our progress since, we now provide a slightly revised version that is similar to Blaha (2004).

The current state of our knowledge of the evolution of the universe has now been extended back in time to about 350,000 years after the Big Bang through recent astrophysical research. While this progress is encouraging we still face major issues: the nature of Dark Matter (hopefully resolved by the Extended Standard Model), the nature and origin of Dark Energy (hopefully resolved in this chapter), and the events of those critical years before the 350,000 year point that we are slowly reaching experimentally. Those early years and the Big Bang itself remain mysteries. This situation is especially critical since the early years of the universe apparently contain an uncertain beginning and an explosive growth.

In this chapter we will attempt to understand that unknown period in the neighborhood of t = 0 where quantum effects we believe play a major role. We will suggest that the inflationary growth of the universe, which is attributed to an unknown "particle", actually is caused by the energy of the q-number part of coordinates – the quantum field $Y^\mu(x)$ that we saw in earlier chapters.

11.3 The State of the Universe at t = 0

If we simply extrapolate the currently popular Standard Cosmological Models (with or without inflations) back to the Big Bang t = 0, we find a universe beginning as a single "mathematical point" with infinite mass density and infinite temperature. The Robertson-Walker metric scale factor a(t), which is a solution of the Einstein equation

$$\dot{a}^2 - 8\pi G\rho a^2/3 = -k \tag{11.1}$$

typically is solved for a perfect fluid under the assumption of a matter-dominated or a radiation-dominated phase of the universe. If we assume the universe is matter-dominated, then the energy density is

Matter-Dominated: $$\rho = \rho_0/a(t)^3 \tag{11.2}$$

Under the alternate assumption that the universe is radiation-dominated we have

Radiation-Dominated: $$\rho = \rho_0/a(t)^4 \tag{11.3}$$

With either assumption we find that the scale factor behaves as

$$a(t) \propto t^n \qquad (11.4)$$

where $0 < n < 1$. Thus $a(0) = 0$ and the universe reduces to a point with infinite density (eqns. 11.2 and 11.3), and with infinite temperature since

$$T \propto a^{-1}(t) \qquad (11.5)$$

There are evidently grave difficulties in extrapolating the Standard Cosmological Model, or its current variants, to t = 0. The difficulty is compounded by the inherently quantum mechanical aspects that are normally associated with gravitation at ultra-small distances.

Currently, the only viable complete theory of Quantum Gravity is the Two Tier Quantum Gravity of Blaha (2003) and (2005a). Blaha's type of quantum field theory has the interesting feature that all forces (particle propagators) become zero at very short distances (presumably much less than the Planck mass). Thus a point universe could have an infinite density of essentially "non-interacting" matter as a quasi-stable state. Furthermore if one uses the *generalized* Robertson-Walker metric as described in Blaha (2004), and Appendix 11-A of this book, one finds a classical scale factor of the form:

$$A(t, \check{r}) = a(t)b(\check{r}) \qquad (11.6)$$

where a(t) satisfies eq. 11.1.

Since quantum effects can be expected to play a role near t = 0 (the Big Bang) it is possible that the expectation value of a quantum scale factor operator, taking account of quantum effects, could have the form

$$\langle A(t, \check{r})\rangle = \langle a(t)\rangle\langle b(\check{r}, t)\rangle \qquad (11.7)$$

where

$$\langle b(\check{r}, t)\rangle \rightarrow \beta(\check{r})/\langle a(t)\rangle \qquad (11.8)$$

as $t \rightarrow 0$ so that the zero of $\langle a(t)\rangle$ might be cancelled with the result

$$\langle A(t, \check{r})\rangle \rightarrow \beta(\check{r}) \neq 0 \qquad (11.9)$$

Quantum effects would thus eliminate the singularities at t = 0. A quantized version of the generalized Robertson-Walker model[111] opens the possibility of a universe with a finite size, density, and temperature at the time of the Big Bang.

---

[111] This model appears in Blaha (2004).

With that possibility in mind, we will use Blaha's (2004) Two Tier theory to develop a version of his quantum model of the universe in the neighborhood of t = 0. Starting from eq. 8.7.1 of that book which is eq. 11-A.7.1 in the following Appendix 11-A:

$$A(t, \check{r}) = a(t)b(\check{r}) = 2\alpha k^{-½}a(t)[1 + \alpha^2 \check{r}^2]^{-1} \quad (8.7.1)$$

where a is a constant given by eq. 11-A.6.3 and introducing a Two Tier variable Y (as in chapter 7 of Blaha (2004)) with the identification[112]

$$\check{r} \equiv M_c X = M_c(y + iY/M_c^2)$$

we see

$$b(\check{r}) = b(y, t) = 2\alpha k^{-½}[1 + \alpha^2(M_c y + iY/M_c)^2]^{-1} \quad (11.10)$$

If

$$Y = -M_c[a_1(y, t) - a_2(y, t)a(t)]^{½}/\alpha + iM_c^2 a_3(y, t) \quad (11.11)$$

and if, as t → 0,

$$a_1(y, t) \to a_1(y, 0) = 1 \quad (11.12)$$

$$a_2(y, t) \to a_2(y, 0) \neq 0 \quad (11.13)$$

$$a_3(y, t) \to a_3(y, 0) = y \quad (11.14)$$

then we find

$$b(y, t) \to 2\alpha k^{-½}/[a_2(y, 0)a(t)] \quad (11.15)$$

and

$$A(t, \check{r}) = a(t)b(\check{r}) = 2\alpha k^{-½}/a_2(y, 0) + a(t)\beta_1(y) + a^2(t)\beta_2(y) + \ldots \quad (11.16)$$
$$\to 2\alpha k^{-½}/a_2(y, 0) \qquad \text{as } t \to 0$$

where we omit symbols indicating expectation values for the sake of clarity. Thus, under these circumstances, space does not collapse to a point and the density and temperature—as well as other parameters of interest—are finite; *and the features of the Standard Cosmological Model at larger times are still valid.*

In our model we will make the following assumptions about the universe near t = 0:

1. The particles in the universe consist of fundamental elementary particles – gravitons, photons, electrons, neutrinos, quarks, gluons and so on – and their corresponding anti-particles.

---

[112] α is a constant and not the fine structure constant.

2. The particles are described by Two Tier quantum field theory. In this type of quantum field theory all particle interactions become negligible at very short distances (as described in chapter 7 of Blaha (2004)) and so the forces between particles may be neglected near $t = 0$ when the universe is immensely *small.*

3. The energy of the universe can be viewed as consisting of particles – bosons and fermions – each species having blackbody energy distributions since the universe is the best of all possible black bodies.

4. The enormous energy of the universe even if confined to a small region makes the classical Einstein equations a good approximation *due to its macroscopic nature* with one proviso (item 5). Therefore we assume the Generalized Robertson-Walker metric of Blaha (2004) and appendix 11-A.

5. In the neighborhood of $t = 0$ when the universe is effectively confined to a region whose scale is set by the Planck mass or smaller the quantum nature of the Two Tier coordinate X^μ becomes significant. In particular the Y^μ field causes a profound change in the behavior of the scale factor A(t, r) as $t \to 0$. (Note: $X^\mu(y) = y^\mu + iY^\mu(y)/M^2$ defines the quantum coordinates where y^μ is a c-number coordinate and Y^μ a free q-number field similar to the electromagnetic field.)

6. The Y^μ quanta are assumed to have a black body spectrum[113] – just like elementary particles – reflecting their continuous emission and absorption by gravitons and other elementary particles. The Y^μ blackbody spectrum is implemented via a coherent state. Effectively the coherent state opens a small "bubble" into complex space-time changing the dynamics of the universe at $t = 0$.

7. We will calculate the expectation value of the quantum field operator Y^μ in a closed Robertson-Walker space. In principle we must use the generalized Robertson-Walker metric since the scale factor will depend on both r and t through its dependence on the expectation value of Y^μ.

[113] The only reasonable choice for the spectrum is a black body spectrum given the confinement of the field to the ultimate black body – the universe.

11.4 Two Tier Quantum Model for the Beginning of the Universe

In our approach in this, and the following, sections we will follow a modest program using the known theoretical foundations of elementary particle physics: the Extendsd Standard Model unified with Quantum Gravity in a Two Tier quantum field theoretic framework. We will supplement this framework with natural assumptions about the initial conditions of the universe in order to develop a theory describing the evolution of the universe from its initial state.

11.4.1 Einstein Equations Near t = 0

There is no physical reason to believe that the universe at the beginning of time, t = 0, was a mathematical point of infinite temperature and density since the extrapolation of the scale factor of the Standard Cosmological Model to t = 0 is unwarranted for many reasons including quantum considerations.

Ideally we would use the Quantum Theory of gravity to establish the physical theory of the universe near t = 0. However a quantum calculation of the global structure of the universe near t = 0 is not feasible. In view of this situation we must find an approximation that captures the physics of the universe near the Big Bang. One approach is based on the macroscopic energy of the early universe. One can expect that a classical gravitation model theory with appropriate quantum corrections may be a reasonable approximation to the early state of the universe. After all, macroscopic bodies are described by classical physics in general. And the universe is a macroscopic body by virtue of its content at the point of the Big Bang despite its small size. Therefore we will assume that we may start with a classical gravitation model and then introduce quantum corrections.

The natural first choices – based on symmetry considerations – are a Robertson-Walker model and a generalized Robertson-Walker model of the type described in chapter 8 of Blaha (2004) and Appendix 11-A following. The quantum part, that we will shortly introduce, will require us to use a generalized Robertson-Walker model since the quantum corrections reduce the symmetry to a maximally symmetric *two-dimensional* subspace within a four-dimensional space-time. *The quantum part eliminates the equivalence of the classical Robertson-Walker and generalized Robertson-Walker models* that was described in section 8.7 of Blaha (2004). See appendix 11-A.

Therefore we begin with the classical c-number equation for the invariant interval defined by

$$d\tau^2 = dt^2 - A^2(t, \check{r})[d\check{r}^2 + \check{r}^2(d\theta^2 + \sin^2\theta\, d\varphi^2)] \qquad (11.17)$$

where

$$A(t, \check{r}) = a(t)b(\check{r}) = 2\alpha k^{-1/2}a(t)[1 + \alpha^2\check{r}^2]^{-1} \qquad (11.18)$$

where a is a constant[114] given by eq. 11-A.6.3 and a(t) is the solution of the Einstein equation:

$$\dot{a}^2(t) - 8\pi G\rho a^2(t)/3 = -k \qquad (11.19)$$

Next we introduce quantum coordinates

$$X^\mu = y^\mu + i\, Y^\mu(y)/M_c^{\,2} \qquad (11.20)$$

We choose the same transverse gauge for Y^μ as we did in chapter 7 of Blaha (2004):

$$\partial Y^i/\partial y^i = 0 \qquad (11.21)$$

$$Y^0 = 0 \qquad (11.22)$$

As a result we make the identification (definition of coordinates)

$$X^0 = y^0 \equiv t \qquad (11.23)$$
$$X^j = y^j + i\, Y^j(y)/M_c^{\,2} \equiv M_c^{-1}\check{r}^j \qquad (11.24)$$

The mass factor on the right side of the equal sign in eq. 11.20 is required on dimensional grounds if y is to have the usual dimension of length (inverse mass). As a result, since $\check{r} \in [0, 1]$ by Blaha (2004), eq. 11.24 implies

$$y = |\mathbf{y}| \in [0, M_c^{-1}] \qquad (11.25)$$

There are two constants with the dimension of mass to a power: k and M_c. The constant k determines the curvature of space – a large-scale feature of Robertson-Walker models. The constant M_c is related to the very short distance behavior of the theory – high energy phenomena with energies of the order of the Planck mass or larger, and, as we will see, the origin of the universe – a short distance, high energy phenomena as well. Therefore we have also chosen to use M_c on the right side of eq. 11.24.

Since X^0 is a c-number and since the density $\rho(t)$ is a large c-number to very good approximation we will assume a(t) is the c-number solution of the classical c-number eq. 11.19 as $t \to 0$. Further we assume that quantum effects appear solely through $b(\check{r})$. We also assume that the q-number equivalent of $b(\check{r})$ is $b(M_cX)$. These assumptions are consistent with applying Ockham's Razor. The function $b(M_cX)$ satisfies the functional equation:[115]

[114] We use a to denote a constant, later set equal to 1, and a(t) to be the solution of the Einstein equation. They are not connected to each other.

[115] See eq. 11-A.5.3 in Appendix 11-A.

$$k + (M_c^2 X b^2)^{-1} \partial(Xb'/b)/\partial X = 0 \tag{11.26}$$

where

$$b' = \partial b/\partial X \tag{11.27}$$

and $X = (\vec{X}\cdot\vec{X})^{1/2}$. The formal solution of eq. 11.26 has the same functional form as the c-number solution $b(\check{r})$ in eq. 11.18. Therefore

$$b(M_c X) = :2a k^{-1/2}[1 + a^2 M_c^2 X^2]^{-1}: \tag{11.28}$$

where a is a constant given by eq. 11-A.6.3 and where we have specified normal ordering with : … : to avoid trivial divergences.

Since eq. 11.28 is a q-number expression we must find the scale factor as the expectation value of A(t, X) for a suitable state. We note that, at this point, the invariant interval is an operator expression of the form:

$$d\tau^2 = dt^2 - B^2(t, X)[dX^2 + X^2(d\theta^2 + \sin^2\theta\, d\varphi^2)] \tag{11.29}$$

where

$$B(t, X) = M_c A(t, M_c X) = a(t) b_M(X) \tag{11.30}$$

and

$$b_M(X) = M_c b(M_c X) = :2a M_c k^{-1/2}[1 + a^2 M_c^2 X^2]^{-1}: \tag{11.31}$$

where a is a constant given by eq. 11-A.6.3. Thus the expectation value of $b_M(X)$ also must be calculated in order to determine the invariant interval's expectation value.

11.4.2 Y Black-Body Coherent States

The Y quanta are continuously being emitted and absorbed by the particles in the primeval universe. As such, they may be expected to have a blackbody energy spectrum that is similar to that of the particles from which they derive their existence. In particular one expects the temperature T associated with their blackbody energy distribution to be the same as that of the "real" particles in the universe. After all, the universe is a black body.

Thus the blackbody energy of Y-quanta as a function of frequency ν per unit volume per unit frequency is assumed to be:

$$u_\nu = 8\pi h c^{-2} \nu^3 [e^{h\nu/\kappa T} - 1]^{-1} \tag{11.32}$$

where c is the speed of light, h is Planck's constant, and κ is Boltzmann's constant. At this point we adopt units in which c = 1 and $\hbar = h/2\pi = 1$.

The Hamiltonian for the Y field has a form that is familiar from electrodynamics

$$H = \int d^3y\, \mathscr{H}_Y(y) = \tfrac{1}{2}\int d^3y\, :E_Y{}^2 + B_Y{}^2: = \int d^3p\, \omega \sum_\lambda a^\dagger(p,\lambda)a(p,\lambda) \qquad (11.33)$$

where E_Y and B_Y are the "electric" and "magnetic" fields of the Y field, $\omega = p^0 = |\vec{p}| = 2\pi\nu$ is the energy (in our units), and λ labels the polarization. Note that we are using "infinite volume" continuum quantization formulation.

We now define coherent Y field bra and ket states that yield a spherically symmetric blackbody distribution as the eigenvalue of the Hamiltonian H:

$$|BB, T\rangle = N \quad \exp[\int d^3p\, f(\omega,T)\sum_\lambda a^\dagger(p, \lambda)]|0\rangle \qquad (11.34)$$

$$\langle BB, T| = N^*\langle 0|\exp[\int d^3p\, f^*(\omega,T)\sum_\lambda a(p, \lambda)] \qquad (11.35)$$

where $\omega = |\vec{p}|$, and where N is a normalization factor. The expectation value of H is

$$\langle BB, T|H|BB, T\rangle = \int d^3p\, 2\omega|f(\omega,T)|^2 \qquad (11.36)$$

$$= \int d\omega\, 8\pi\omega^3|f(\omega,T)|^2 \qquad (11.37)$$

$$= \int d\nu\, 16\pi^2\omega^3|f(\omega,T)|^2 \qquad (11.38)$$

where $\omega = 2\pi\nu$ in our units ($c = 1$, $\hbar = 1$), and where the factor of two in eq. 11.36 is the number of polarizations.

The expectation value (eigenvalue) of the energy per unit frequency is

$$H_\nu = 16\pi^2\omega^3|f(\omega,T)|^2 \qquad (11.39)$$

We relate H_ν to the blackbody energy *per unit volume* per unit frequency u_ν using

$$H_\nu = u_\nu(2\pi/\omega)^3 \qquad (11.40)$$

where the factor of $(2\pi/\omega)^3$ makes the right side of eq. 11.40 the blackbody energy per unit frequency in the continuum case of a quantum field in a space of infinite volume. Thus we find

$$f(\omega,T) = \omega^{-3/2}[e^{\omega/\kappa T} - 1]^{-1/2} \qquad (11.41)$$

with the phase of $f(\omega,T)$ set to zero.

11.4.3 Expectation Value of Y in Coherent States

As a preliminary to the evaluation of the operator scale factor in eq. 11.31 we will evaluate the expectation value of powers of the Y field between black body coherent states defined by eqns. 11.34-11.35. We will then determine the expectation value of $b_M(X)$ in combination with a(t) to obtain the behavior of the overall scale factor near t = 0. It should be apparent to the reader that the expectation value of $b_M(X)$ is dependent on t as well as y due to the time dependence of the Y field. Thus the scale factor will exhibit a considerably more intricate behavior than simply its a(t) dependence.

The Fourier expansion of the Y field is:

$$Y^i(z) = \int d^3p\, N_0(\omega) \sum_{\lambda=1}^{2} \varepsilon^i(p, \lambda)[a(p,\lambda)\, e^{-ip\cdot z} + a^\dagger(p,\lambda)\, e^{ip\cdot z}] \qquad (11.42)$$

where z^μ will be set equal to y^μ later, and where

$$N_0(\omega) = [(2\pi)^3 2\omega]^{-½} \qquad (11.43)$$

and

$$\omega = (\mathbf{p}^2)^{½} = p^0 \qquad (11.44)$$

with $\vec{\boldsymbol{\varepsilon}}(p, \lambda)$ being the polarization unit vectors for $\lambda = 1, 2$ and $\eta_{\mu\nu}p^\mu p^\nu = 0$. The expectation value of Y between the |BB, T> states is:

$$<BB, T|Y^i(z)|BB, T> = \int d^3p\, N_0(\omega)f(\omega,T)[e^{-ip\cdot z} + e^{ip\cdot z}] \Sigma\varepsilon^i(p, \lambda) \qquad (11.45)$$

The evaluation of eq. 11.45 (and spherical symmetry) gives

$$<BB, T|Y^i(z)|BB, T> = \hat{y}^i \int d^3p\, N_0(\omega)f(\omega,T)[e^{-ip\cdot z} + e^{ip\cdot z}] \sum_\lambda \hat{\mathbf{z}}\cdot \boldsymbol{\varepsilon}(p, \lambda) \qquad (11.46)$$
$$\equiv \hat{\mathbf{z}}^i\, Y_{BB}(t, z)$$

where $\hat{\mathbf{z}} = \vec{\mathbf{z}}/|\vec{\mathbf{z}}|$ is the unit three-vector in the direction of $\vec{\mathbf{z}}$, $z = |\vec{\mathbf{z}}|$, and $p\cdot z = \omega(t - z\cos\theta)$. We define a spatial coordinate system – choosing the z-axis parallel to $\vec{z}$. Then we have

$$\vec{z} = (0, 0, z) \qquad (11.47)$$
$$\vec{p} = (\sin\theta\cos\phi\, ,\, \sin\theta\sin\phi,\, \cos\theta) \qquad (11.48)$$
$$\vec{\boldsymbol{\varepsilon}}(p,1) = (\cos\theta\cos\phi\, ,\, \cos\theta\sin\phi,\, -\sin\theta) \qquad (11.49)$$
$$\vec{\boldsymbol{\varepsilon}}(p,2) = (-\sin\phi\, ,\, \cos\phi,\, 0) \qquad (11.50)$$

with the result (taking account of eq. 11.46)

$$Y_{BB}(t, z) = \langle BB, T|\, \hat{\mathbf{z}}\cdot \mathbf{Y}(t, z)\, |BB, T\rangle$$

$$= 2\pi\int_0^\infty d\omega\, \omega^2 N_0(\omega) f(\omega,T)\int_0^\pi d\theta\, \sin^2\theta\, [e^{-ip\cdot z} + e^{ip\cdot z}] \quad (11.51)$$

where $p\cdot z = \omega(t - z\cos\theta)$ with $z = |\vec{z}|$. We will develop integral representations and approximations to Y_{BB} in a later section.

11.4.4 Expectation Values of the Scale Factor A(t, X) and the Invariant Interval $d\tau^2$

The scale factor

$$b_M(X) = :2aM_ck^{-\frac{1}{2}}[1 + a^2M_c^2X^2]^{-1}: \quad (11.52)$$

where a is a constant given by eq. 11-A.6.3, is a normal-ordered q-number expression. We can formally expand (define) this expression as a power series of normal-ordered powers of X^2 and then evaluate it between blackbody coherent states. First we note that

$$\langle BB, T|:Y^{i1}(z)Y^{i2}(z)Y^{i3}(z)Y^{i4}(z) \ldots Y^{in}(z):|BB, T\rangle = \hat{z}^{i1}\hat{z}^{i2}\hat{z}^{i3}\hat{z}^{i4} \ldots \hat{z}^{in}(Y_{BB}(t, z))^n \quad (11.53)$$

We now set $z^\mu = y^\mu$, and use Y_{BB} to represent $Y_{BB}(t, \mathbf{y})$:

$$Y_{BB} \equiv Y_{BB}(t, \vec{y}) \quad (11.54)$$

Thus

$$b_{BB}(y, t) = \langle BB, T|b_M(X)|BB, T\rangle$$

$$= 2aM_ck^{-\frac{1}{2}}\{1 + a^2[M_c^2y^2 + 2i\, yY_{BB} - Y_{BB}^2/M_c^2]\}^{-1} \quad (11.55)$$

and

$$B_{BB}(t, y) = \langle BB, T|B(t, X)|BB, T\rangle = a(t)b_{BB}(y, t) \quad (11.56)$$

where a is a constant given by eq. 11-A.6.3. The expectation value of the q-number invariant interval (eq. 11.29) is the c-number expression:

$$d\tau_{BB}^2 = \langle BB, T|d\tau^2|BB, T\rangle$$

$$= dt^2 - B_{BB}^2(t, y)[dX_{BB}^2 + X_{BB}^2(d\theta^2 + \sin^2\theta\, d\varphi^2)] \quad (11.57)$$

where

$$X_{BB} = y + iY_{BB}/M_c^2 \quad (11.58)$$

and

$$dX_{BB} = dy(1 + iM_c^{-2}\partial Y_{BB}/\partial y) \quad (11.59)$$

The appearance of Y_{BB} in the expression for the invariant interval (eq. 11.57) has two effects: it introduces complex space-time into the model and the generalized Robertson-Walker metric is no longer equivalent to the Robertson-Walker metric for all a as it would be in the classical case.

We will see that Y_{BB} approaches zero at large times thus yielding the conventional Robertson-Walker models. But at small times of the order of the Planck time near the Big Bang we enter a brave new world of complex space-time. We will investigate the nature of this new complex world in the succeeding sections.

11.4.5 Representation and Approximations for $Y_{BB}(t, z)$

The angle integral in eq. 11.51 can be performed to yield

$$Y_{BB}(t, z) = \pi^{½}z^{-1} \int_0^\infty d\omega\, \omega^{-1}(e^{\omega/\kappa T} - 1)^{-½}\cos(\omega t)J_1(\omega z) \qquad (11.60)$$

where $J_1(\omega z)$ is a Bessel function using 3.915.5 of Gradshteyn (1965).

11.4.5.1 Some Representations of Y_{BB}

The integral in eq. 11.60 does not appear to be simply expressible in terms of standard transcendental functions. A series representation of the integral can be obtained by expanding the exponential factor due to the Planck distribution:

$$Y_{BB}(t, z) = ½\pi^{½}\kappa T \sum_{n=0}^{\infty} (2n)![2^{2n}(n!)^2]^{-1}\{(2n+1+ 2i\kappa Tt)^{-1} F(½, 1; 2; -[\kappa Tz/(n+½ + i\kappa Tt)]^2) + (2n+1 - 2i\kappa Tt)^{-1} F(½, 1; 2; -[\kappa Tz/(n+½ - i\kappa Tt)]^2)\} \qquad (11.61)$$

where $F(a, b; c; w)$ is a hypergeometric function.[116]

Using an integral representation[117]

$$F(a, b; c; w) = \Gamma(c)[\Gamma(b)\Gamma(c - b)]^{-1} \int_0^1 dt\, t^{b-1}(1 - t)^{c-b-1}(1 - tw)^{-a}$$

for $F(½, 1; 2; w)$ we see eq. 11.61 can be written in terms of simpler algebraic expressions:

$$Y_{BB}(t, z) = ½\pi^{½}(\kappa Tz^2)^{-1} \sum_{n=0}^{\infty}(2n)![2^{2n}(n!)^2]^{-1}\{[(n+ ½ + i\kappa Tt)^2 + (\kappa Tz)^2]^{½} +$$

---

[116] Based on the integral 6.613.1 on p. 711 of Gradshteyn (1965).

[117] Magnus (1949) p. 8.

$$+ [(n+ \tfrac{1}{2} - i\kappa Tt)^2 + (\kappa Tz)^2]^{1/2} - (2n+1)\} \tag{11.62}$$

Eq. 11.62 shows the limit of $Y_{BB}(t, z)$ for large t is

$$Y_{BB}(t, z) \to \pi^{1/2}\kappa T \sum_{n=0}^{\infty}(2n)![2^{2n+2}(n!)^2]^{-1}(2n + 1)[(n + \tfrac{1}{2})^2 + (\kappa Tt)^2]^{-1} \to 0 \tag{11.63}$$

if $tT \to \infty$ as $t \to \infty$ as we see in cosmological models (see section 11.4).

Thus

$$(b_{BB}(y, t))^2 dX^2 \to 2aM_c^2k^{-1}[1 + a^2M_c^2y^2]^{-2}dy^2 \equiv 2ak^{-1}[1 + a^2\check{r}^2]^{-2}d\check{r}^2 \tag{11.64}$$

where a is a constant given by eq. 11-A.6.3, for large t showing the Two Tier cosmological model becomes a Robertson-Walker model at large times. However, the Two Tier standard cosmological model is very different at small times of the order of the Planck time near the Big Bang point.

11.4.5.2 Approximate Solution for Y_{BB}

The integral representation and power series representation of $Y_{BB}(t, y)$ do not reveal the physical behavior of the model for small times t and distances y. Therefore we will examine an approximation for $Y_{BB}(t, y)$ for ranges of y, t and T that are relevant for our considerations. We begin by scaling the integration variable in eq. 11.60 with the result:

$$Y_{BB}(t, y) = \pi^{1/2}y^{-1}\int_0^{\infty} d\omega\, \omega^{-1}(e^{\omega} - 1)^{-1/2}\cos(\omega\kappa Tt)J_1(\omega\kappa Ty) \tag{11.65}$$

The blackbody exponential factor $(e^{\omega} - 1)^{-1/2}$ in the integrand of $Y_{BB}(t, y)$ enables the leading order approximate behavior of $Y_{BB}(t, y)$ to be determined for $0 \le t \lessapprox 10^{108}$ s – for all time, practically speaking. In a later section (section 11.3.4) we will see that our approximation to the integral in eq. 11.65 is consistent with the solution that we obtain for Y_{BB}, for the scale factor and thus for the temperature T. The approximations that we will make in eq. 11.65 are

$$\cos(\omega\kappa Tt) \approx 1 \tag{11.66}$$

$$J_1(\omega\kappa Ty) \approx \omega\kappa Ty/2 \tag{11.67}$$

They are based on $\kappa Tt \ll 1$ and $\kappa Ty \ll 1$ for all y ($0 \le y \le M_c^{-1}$). The exponential factor tends to limit contributions to the integral to small ω. After making these approximations we find

$$Y_{BB}(t, y) \simeq \tfrac{1}{2}\,\pi^{1/2}\kappa T \int_0^\infty d\omega\, (e^{\omega} - 1)^{-1/2}$$

$$\simeq \pi^{3/2}\kappa T/2 \qquad (11.68)$$

The limit as t gets large can also be approximately determined from eq. 11.65. For large t such that κTy is small (and ω is small due to the exponential Planck distribution factor) we can again approximate the Bessel function with its leading power series expansion term and the exponential factor can again be approximated by $e^{\omega} - 1 \approx \omega$ so that eq. 11.65 becomes approximately

As $t \to \infty$:

$$Y_{BB}(t, y) \simeq \pi^{1/2}2^{-1}\,\kappa T \int_0^\infty d\omega\, \omega^{-1/2}\cos(\omega\kappa Tt)$$

$$= \pi 2^{-3/2}[\kappa T/t]^{1/2} \qquad (11.69)$$

using 3.751.2 of Gradshteyn (1965). We note that, while κTy is small, κTt could possibly have been large in either a matter-dominated or radiation-dominated universe since it grows as t to a positive power (see section 11.3.) *However, since $Y_{BB}(t, y)$ approaches zero for large times its impact can only be seen in the initial formative stages of the universe near t = 0.*

11.5 The Scale Factor a(t) Near t = 0

The "time factor" a(t) of the scale factor $B_{BB}(t, y)$ appears in

$$B_{BB}(t, y) = \langle BB, T|B(t, X)|BB, T\rangle = a(t)b_{BB}(y, t) \qquad (11.70)$$

and is determined by the classical Einstein equation:

$$\dot{a}^2(t) - 8\pi G\rho a^2(t)/3 = -k \qquad (11.71)$$

As we have argued earlier, the source determining a(t) for small times in the neighborhood of t = 0 (the time of the Big Bang) is a large, macroscopic, classical density ρ(t) and thus a(t) may be considered to be a c-number quantity determined by the c-number Einstein equation to good approximation. This approximation should continue to hold even if this macroscopic density becomes enormous as $t \to 0$. The quantum effects near t = 0 in the Two Tier model, that we have developed, appear in the factor $b_{BB}(y, t)$ that we evaluated in previous sections.

11.6 A Complex Blackbody Temperature Near t = 0

The blackbody temperature T for relativistic particles (presumably the dominant type of particles near t = 0) is inversely proportional to the scale factor. At large times the blackbody temperature has the form

$$T = T_0/a(t) \tag{11.72}$$

where T_0 is a constant.

At times in the neighborhood of t = 0 (the Big Bang) space has three complex dimensions in the Two Tier model. Temperature can be viewed as a measure of the root mean square speed (or the "average energy") of the components of the perfect fluid that we have assumed. In the case of a gas of particles of average energy E:

$$T = E/(3k/2) \tag{11.73}$$

In a complex space it is quite natural for the root mean squared speed to be complex as well. As a result complex temperatures naturally follow. Thus we will define

$$T = T_0/B_{BB}(t, y) \tag{11.74}$$

for all time since t = 0. Since $B_{BB}(t, y)$ approaches $M_c a(t)b(\check{r})$ at large times we find its large time behavior is consistent with those of standard Robertson-Walker models. At times near t = 0, the blackbody temperature T is complex since space is complex and complex kinetic energy is allowed. Then we apply a Reality group transformation to obtain the physical temperature.

In the case of the complex temperature T above, the corresponding physical temperature is its absolute value (obtained by multiplying T by a phase factor from the Reality group)

$$T_{physical} = |T_0/B_{BB}(t, y)| \tag{11.74a}$$

We shall use eq. 11.74 for the temperature, transforming the results below, afterwards, to real values using the 4-dimensional Reality group.

11.7 The Nature of the Universe Near t = 0

At this point we are ready to examine the Two Tier model for the Big Bang period.

11.7.1 Behavior of the Complete Scale Factor B(t, y) Near t = 0

The behavior of the expectation value of the scale factor B(t, y), under the assumption that the Y quanta have a blackbody spectrum, is described by the equations:

$$b_{BB}(y, t) = 2\alpha M_c k^{-1/2}\{1 + \alpha^2[M_c^2 y^2 + 2i\, yY_{BB} - Y_{BB}^2/M_c^2]\}^{-1} \tag{11.75}$$

$$B_{BB}(t, y) = \langle BB, T|B(t, X)|BB, T\rangle = a(t)b_{BB}(y, t) \tag{11.76}$$

where a is a constant given by eq. 11-A.6.3, and

$$Y_{BB}(t, y) \simeq \pi^{3/2}\kappa T/2 \tag{11.77}$$
$$a(t) = [2\pi G\rho_0 n^2/3]^{1/n} t^{2/n} \tag{11.78}$$
$$T = T_0/B_{BB}(t, y) \tag{11.79}$$

as $t \to 0$. We will set the constant $\alpha = 1$ in the interests of simplicity knowing that this value results in a metric fully equivalent to the Robertson-Walker metric at large times. (Other values of α[118] would also result in a metric equivalent to the Robertson-Walker metric at large times after a re-scaling of the radial coordinate.) Thus we may write

$$B_{BB}(t, y) \cong 2k^{-1/2}M_c a(t)\{1 + M_c^2y^2 + iy\pi^{3/2}\kappa T_0/B_{BB} - \pi^3\kappa^2T_0^2/(4M_c^2B_{BB}^2)\}^{-1} \tag{11.80}$$

This quadratic algebraic equation for B_{BB} has the solutions:

$$B_{BB}(t, y) \cong (1 + M_c^2y^2)^{-1}\{-i\varkappa M_c y + k^{-1/2}M_c a(t) \pm [\varkappa^2 - 2i\varkappa yk^{-1/2}M_c^2a(t) + k^{-1}M_c^2a^2(t)]^{1/2}\} \tag{11.81}$$

with

$$\varkappa = \pi^{3/2}\kappa T_0/(2M_c) \tag{11.82}$$

As t gets very large we obtain the equivalent of the Robertson-Walker metric scale factor in this approximation if we choose the plus sign in eq. 11.81 (assuming $a^2(t)$ becomes very large so other terms within the square root can be neglected):

$$B_{BB}(t, y) \to 2k^{-1/2}M_c a(t)/(1 + M_c^2y^2) \tag{11.83}$$

Thus we must choose the plus sign in eq. 11.81:

$$B_{BB}(t, y) \cong (1 + M_c^2y^2)^{-1}\{-i\varkappa M_c y + k^{-1/2}M_c a(t) + [\varkappa^2 - 2i\varkappa yk^{-1/2}M_c^2a(t) + k^{-1}M_c^2a^2(t)]^{1/2}\} \tag{11.84}$$

At t = 0 (the Big Bang) eq. 11.84 simplifies to (assuming a(0) = 0)

$$B_{BB}(0, y) \cong [\varkappa - i\varkappa M_c y]/(1 + M_c^2y^2) \tag{11.85}$$

[118] It is **not** the fine structure constant.

For small y, the real part of $B_{BB}(0, y)$ is a constant and the imaginary part of $B_{BB}(0, y)$ is proportional to y.

11.7.2 The Expectation Value of the Scale Factor $A_{BB}(0, y)$ near t = 0

Eq. 11.85 gives the approximate behavior of the expectation value of the scale factor $B_{BB}(0, y)$ near t = 0 as a function of y. If we compare eq. 11.85 with the mechanism described in eqns. 11.1.6 – 11.1.9 of section 11.1 for cancelling the a(t) factor within the complete scale factor we see that we have found the blackbody spectrum of the Y quanta implements this mechanism. The solution can be written in the form:

$$A_{BB}(t, y) = M_c^{-1}B_{BB}(t, y) \cong \beta_0(y) + \beta_1(y)a(t) + \ldots \quad (11.86)$$

$$\beta_0(y) = \varkappa(1 - iM_cy)/[M_c(1 + M_c^2y^2)] \quad (11.87)$$

$$\beta_1(y) = k^{-1/2}(1 - iM_cy)/(1 + M_c^2y^2) \quad (11.88)$$

Eqns. 11.86–11.88 are expressed in terms of the y variable. They can be expressed in terms of ř as:

$$A(t, \check{r}) = a(t)b(\check{r}) = 2\alpha k^{-1/2}a(t)[1 + \alpha^2\check{r}^2]^{-1} \quad (11.89)$$

where a is a constant given by eq. 11-A.6.3 set to $\alpha = 1$. Evidently, we have $M_cy \equiv \check{r}$ at the level of approximation that we are using. Furthermore we can use

$$\check{r} = \{[1 - (1 - kr^2)^{1/2}]/[1 + (1 - kr^2)^{1/2}]\}^{1/2} \quad (11.90)$$

to express ř in terms of the Roberson-Walker radial coordinate r (from Appendix 11-A.) Thus we find that the Robertson-Walker scale factor a(t) becomes

$$a(t) \to (1 + M_c^2y^2)(2k^{-1/2})^{-1}A_{BB}(t, y) \equiv a_{BBRW}(t, \check{r}) \quad (11.91)$$

$$a_{BBRW}(t, \check{r}) \cong \beta_{0RW}(\check{r}) + \beta_{1RW}(\check{r})a(t) + \ldots \quad (11.92)$$

using the subscript "RW" to denote quantities scaled to the standard Robertson-Walker metric, with

$$\beta_{0RW}(\check{r}) = \varkappa(1 - i\check{r})/[2k^{-1/2}M_c] \quad (11.93)$$

$$\beta_{1RW}(\check{r}) = (1 - i\check{r})/2 \quad (11.94)$$

where ř is specified by eq. 11.90.

Using the Reality group we can "rotate" the complex scale factor, which is a factor in coordinate expressions, to a real value—its absolute value

$$a(t) \rightarrow (1 + M_c^2 y^2)(2k^{-1/2})^{-1}|A_{BB}(t, y)| \equiv |a_{BBRW}(t, \check{r})| \qquad (11.91a)$$

We will study the implications of this scale factor in the following chapters.

Appendix 11-A. Derivation of the Extended Robertson-Walker Model

This appendix provides the derivation of a generalization of the Roberson-Walker solution of General Relativity that we used in chapter 11 in our discussion of the Quantum Big Bang Theory originally presented in Blaha (2004). This appendix is extracted from chapter 8 of Blaha (2004) as necessary background information for chapter 11 and the following chapters.

The generalization derived here has not been derived before, to our knowledge, because at the level of c-number General Relativity it is fully equivalent to the known Robertson-Walker model. However when the theory has q-number parts introduced in a physically meaningful way, as we do, it is no longer equivalent to the c-number Robertson Walker model.

11-A.1 The Robertson-Walker Metric

Much of the current modeling of the evolution and properties of the universe is based on the assumption of a Robertson-Walker metric which is used in the Einstein equations to obtain a first order differential equation for the scale factor R(t):

$$\dot{R}^2 + k = 8\pi G\rho R^2/3 \qquad (11\text{-A.1.1})$$

where k is a factor in the three-dimensional spatial curvature of the Robertson-Walker metric:

$$K_3(t) = k/R^2(t) \qquad (11\text{-A.1.2})$$

Eq. 11-A.1.2 suggests that accurate measurements of the Hubble constant and other cosmological quantities could lead to an accurate determination of the curvature, the time dependence of the Hubble constant, and of R(t).

The form of the Robertson-Walker metric (We consider a real space-time only in this appendix) follows from the assumption of a maximally symmetric three-dimensional subspace whose metric has eigenvalues of the same sign (negative in our formalism) residing within a four-dimensional space-time with one positive eigenvalue and three negative eigenvalues. A maximally symmetric space is isotropic and homogeneous.[119]

---

[119] See Chapter 13 of Weinberg (1972) for a detailed discussion.

Although the Robertson-Walker metric does not appear to embody the concept of an absolute space-time the general arguments presented in chapter 2 of Blaha (2004) indicate that it does in fact implicitly define an absolute reference frame.[120]

Therefore it is sensible to inquire whether a more general metric – a generalization of the Robertson-Walker metric – might be worth investigating – particularly in view of the major unexplained mysteries of Dark Energy as well as other new data showing the existence of massive black holes and quite mature galaxies shortly after (two or three billion years) the origin of the universe. The pile-up of mysteries from WMAP, SDSS and other sources indicates a reconsideration of fundamental assumptions may be worthwhile.

Therefore we will examine the "simplest" generalization of the Robertson-Walker metric in this appendix with a view towards elucidating some of these mysteries. More importantly, we will use this generalization in chapter 19 and subsequent chapters to develop a non-singular, Two Tier formulation of the dynamics of the universe "at the beginning of time" – the Big Bang – taking account of quantum effects.

From the point of view of the definition of maximally symmetric subspaces the most immediate generalization of the Robertson-Walker metric is to assume a maximally symmetric *two-dimensional* subspace within a four-dimensional space-time. The general form of the metric in this case is:

$$d\tau^2 = A_{tt}(r,t)\, dt^2 + 2A_{rt}(r,t)\, dt\, dr + A_{rr}(r,t)\, dr^2 + B(r,t)(d\theta^2 + \sin^2\theta d\varphi^2) \quad (11\text{-A.1.3})$$

where A_{ik} is a 2×2 symmetric matrix with one positive and one negative eigenvalue, and B(r,t) is a negative function of r and t.

11-A.2 A Generalization of the Roberson-Walker Metric

We shall consider a generalization of the Robertson-Walker metric (eq. 5.5.1 of Blaha (2004)), which is a special case of eq. 11-A.1.3 that preserves the overall form of the Robertson-Walker metric but allows the scale factor a(t) to depend on r as well as t:

$$R(t) \rightarrow A_0(t, r) \quad (11\text{-A.2.1})$$

The generalized metric that we will analyze is embodied in the invariant interval expression:

$$d\tau^2 = dt^2 - A_0^2(t, r)[dr^2/(1 - kr^2) + r^2(d\theta^2 + \sin^2\theta\, d\varphi^2)] \quad (11\text{-A.2.2})$$

[120120] This conclusion is now obvious due to the existence of the Megaverse. We note there can only be one absolute reference frame up to a Lorentz transformation since the sets of inertial reference frames of two absolute reference frames must be the same. This implies any two absolute reference frames must be related by a Lorentz transformation since absolute frames are necessarily flat inertial frames.

The introduction of a dependence on the radius r in $A_0(t, r)$ in eq. 11-A.2.2 eliminates the homogeneity of the three-dimensional spatial subspace reducing it to a two-dimensional maximally symmetric subspace.

We note that the general solution of the Einstein equations for the standard case of a perfect fluid lead to the usual view of the expansion of the universe (after the Big Bang Epoch), Hubble's law, the red shifts of radiation from distant sources of radiation, and the Cosmic Microwave Background (CMB) radiation.

11-A.3 The Einstein Equations for the Generalized Robertson-Walker Metric

The Einstein equations can be written

$$R_{\mu\nu} = -8\pi G S_{\mu\nu} \quad \text{(11-A.3.1a)}$$

$$S_{\mu\nu} = T_{\mu\nu} - \tfrac{1}{2} g_{\mu\nu} T^{\sigma}{}_{\sigma} \quad \text{(11-A.3.1b)}$$

where $R_{\mu\nu}$ is the Ricci tensor and $T_{\mu\nu}$ is the energy-momentum tensor. Assuming the energy-momentum tensor has the form of the energy-momentum tensor of a perfect fluid with the only non-zero components:

$$T_{tt} = \rho\, g_{tt} \quad \text{(11-A.3.2)}$$

$$T_{rr} = -\, p g_{rr} \quad \text{(11-A.3.3)}$$

$$T_{\theta\theta} = -\, p g_{\theta\theta} \quad \text{(11-A.3.4)}$$

$$T_{\phi\phi} = -\, p g_{\phi\phi} \quad \text{(11-A.3.5)}$$

where ρ is the density and p is the pressure, then the non-zero components of $S_{\mu\nu}$ are:

$$S_{tt} = \tfrac{1}{2}\,(\rho + 3p)\, g_{tt} \quad \text{(11-A.3.6)}$$

$$S_{rr} = -\tfrac{1}{2}\,(\rho - p) g_{rr} \quad \text{(11-A.3.7)}$$

$$S_{\theta\theta} = -\tfrac{1}{2}\,(\rho - p) g_{\theta\theta} \quad \text{(11-A.3.8)}$$

$$S_{\phi\phi} = -\tfrac{1}{2}\,(\rho - p) g_{\phi\phi} \quad \text{(11-A.3.9)}$$

In particular, the fact that

$$S_{tr} = 0 \quad \text{(11-A.3.10)}$$

for a perfect fluid results in an important simplification in the solution of the Einstein equations for this case.

The density $\rho = \rho(t)$ and the pressure $p = p(t)$ are assumed to be solely functions of time t as is usual in the case of a perfect fluid.

11-A.4 The Differential Equations for the Generalized Scale Factor A(t, r)

The dependence of the scale factor A_0 on both r and t leads to a significantly more complicated calculation of the Ricci tensor. We start by noting

$$g_{tt} = 1 \qquad g_{rr} = -A_0^2/(1 - kr^2) \qquad g_{\theta\theta} = -A_0^2r^2 \qquad g_{\phi\phi} = -A_0^2r^2\sin^2\theta \tag{11-A.4.1}$$

Despite the dependence of A_0 on both t and r in the generalized case a direct calculation of the tt Ricci tensor component yields the familiar expression:

$$R_{tt} = g_{tt}\, 3\ddot{A}_0/A_0 \tag{11-A.4.2}$$

where we use dots over A_0 to indicate partial derivatives with respect to time:

$$\ddot{A}_0 \equiv \partial^2 A_0/\partial t^2 \tag{11-A.4.3}$$

However, the tr-component of the Ricci tensor R_{tr}, which is zero in the case of the ordinary Robertson-Walker metric, is non-zero in the more general case under consideration:

$$R_{tr} = 2\, \partial(\partial A_0/\partial t)/\partial r \tag{11-A.4.4}$$

The corresponding Einstein equation is

$$R_{tr} = 2\, \partial(A_0^{-1}\, \partial A_0/\partial t)/\partial r = -8\pi G S_{tr} = 0 \tag{11-A.4.5}$$

for a perfect fluid. Eq. 11-A.4.5 implies that $A_0(t, r)$ factorizes:

$$A_0(t, r) = a(t)b_0(r) \tag{11-A.4.6}$$

This factorization results in a substantial simplification in the non-linear Einstein equations considered next, which are shown to be separable in the radial and time variables.

Before proceeding to the consideration of the remaining Einstein equations it is convenient to redefine the radial coordinate using

$$\check{r}\, b(\check{r}) = r\, b_0(r) \tag{11-A.4.7}$$

and

$$d\check{r}/dr = b_0(r)[b(\check{r})(1 - kr^2)^{1/2}]^{-1} = \check{r}\, [r(1 - kr^2)^{1/2}]^{-1} \tag{11-A.4.8}$$

While this change of coordinates does not change the physical content of the theory it does lead to simpler Einstein equations. The change of radial coordinate results in a new form of the invariant interval:

$$d\tau^2 = dt^2 - a^2(t)b^2(\check{r})[d\check{r}^2 + \check{r}^2(d\theta^2 + \sin^2\theta\, d\varphi^2)] \qquad (11\text{-A.4.9})$$

The new radial coordinate ř is related to the old radial coordinate by

$$\check{r} = \{[1 - (1 - kr^2)^{1/2}]/[1 + (1 - kr^2)^{1/2}]\}^{1/2} \qquad (11\text{-A.4.10})$$

and

$$r = 2k^{-1/2}\check{r}(1 + \check{r}^2)^{-1} \qquad (11\text{-A.4.11})$$

Note the range of ř is [0, 1].

A direct calculation of the Ricci tensor for the metric in eq. 11-A.4.9 with A(t, ř) defined as:

$$A \equiv A(t, \check{r}) = a(t)b(\check{r}) \qquad (11\text{-A.4.12})$$

leads to the following Einstein equations (remembering our flat space cartesian metric is $\eta_{tt} = +1$ and $\eta_{ij} = -\delta_{ij}$ for i, j = spatial indices):

$$R_{tt} = g_{tt}3\ddot{A}/A = -8\pi G S_{tt} = -4\pi G(\rho + 3p) \qquad (11\text{-A.4.13})$$

$$R_{t\check{r}} = 2\, \partial(A^{-1}\, \partial A/\partial t)/\partial\check{r} = -8\pi G\, S_{t\check{r}} = 0 \qquad (11\text{-A.4.14})$$

$$R_{\check{r}\check{r}} = g_{\check{r}\check{r}}[\ddot{A}/A + 2(\dot{A}/A)^2 - 2(\check{r}A^2)^{-1}\, \partial(\check{r}\, A'/A)/\partial\check{r}]$$
$$= -8\pi G\, S_{\check{r}\check{r}} = 4\pi G\, g_{\check{r}\check{r}}(\rho - p) \qquad (11\text{-A.4.15})$$

$$R_{\theta\theta} = g_{\theta\theta}[\ddot{A}/A + 2(\dot{A}/A)^2 - A''/A^3 - 3A'/(\check{r}A^3)^{-1}]$$

$$= -8\pi G\, S_{\theta\theta} = 4\pi G\, g_{\theta\theta}(\rho - p) \qquad (11\text{-A.4.16})$$

$$R_{\phi\phi} = g_{\phi\phi}[\ddot{A}/A + 2(\dot{A}/A)^2 - A''/A^3 - 3A'/(\check{r}A^3)^{-1}]$$

$$= -8\pi G\, S_{\phi\phi} = 4\pi G\, g_{\phi\phi}\, (\rho - p) \qquad (11\text{-A.4.17})$$

where

$$A' = \partial A/\partial\check{r} \qquad (11\text{-A.4.18})$$

and

$$A'' = \partial^2 A/\partial\check{r}^2 \qquad (11\text{-A.4.19})$$

11-A.5 The Solution for the Generalized Scale Factor A(t, r)

We begin by substituting eq. 11-A.4.13 in eq. 11-A.4.15, and then substituting the factorization of A (eq. 11-A.4.12). The result is a separable equation:

$$\dot{a}^2 - 8\pi G\rho a^2/3 - (\check{r}b^2)^{-1}\,\partial(\check{r}\,b'/b)/\partial\check{r} = 0 \qquad (11\text{-A.5.1})$$

Since the first two terms in eq. 11-A.5.1 are solely functions of t while the third term is solely a function of $\check{r}$ we obtain the separated equations:

$$\dot{a}^2(t) - 8\pi G\rho a^2(t)/3 = -k \qquad (11\text{-A.5.2})$$

and

$$k + (\check{r}b^2)^{-1}\,\partial(\check{r}\,b'/b)/\partial\check{r} = 0 \qquad (11\text{-A.5.3})$$

where k is a separation constant which we provisionally identify with the curvature parameter k in eq. 11-A.2.2.

Eq. 11-A.5.2 is precisely the equation used for the time dependent scale factor in current cosmological models (eq. 5A.3.2) using the Robertson-Walker metric. Therefore its solution, which depends on the time dependence of the energy density, will be the same as that of the corresponding conventional cosmological model for the same density.

Eq. 11-A.5.3 is a differential equation for the *spatial* expansion scale factor $b(\check{r})$. Under the assumption of a perfect fluid it depends solely on the constant k and is independent of the details of the perfect fluid (i.e. its density and pressure). There are two solutions of the second order non-linear differential equation for $b(\check{r})$. These solutions can be written as

$$b_1(\check{r}) = 2\gamma\delta k^{-1/2}[\delta^2 + \gamma^2\,\check{r}^2]^{-1} \qquad (11\text{-A.5.4})$$

where γ and δ are constants, and

$$b_2(\check{r}) = \sigma k^{-1/2}[\check{r}\,(\varsigma \pm i\sigma \ln \check{r})]^{-1} \qquad (11\text{-A.5.5})$$

where σ and ς are constants.

$b_1(\check{r})$ (in eq. 11-A.5.4) is the only physically acceptable solution for the case of a perfect fluid with energy-momentum tensor specified by eqns. 11-A.3.2 – 11-A.3.5. Reason: The solution $b_1(\check{r})$ satisfies eqns. 11-A.4.16 and 11-A.4.17 while the other solution $b_2(\check{r})$ does not satisfy these equations. A necessary and sufficient condition for $b(\check{r})$ to satisfy eqns. 11-A.4.16 and 11-A.4.17 is that

$$b'' - \check{r}^{-1}\,b' - 2\,b'^2/b = 0 \qquad (11\text{-A.5.6})$$

where ′ denotes a derivative with respect to ř. Eq. 11-A.5.6 follows from subtracting the coefficients of the metric tensor component factors in eq. 11-A.4.15 from the coefficients of the metric tensor component factors in eq. 11-A.4.16 (or eq. 11-A.4.17). The solution $b_1(\check{r})$ satisfies eq. 11-A.5.6 for all values of γ and δ. The solution $b_2(\check{r})$ does not satisfy eq. 11-A.5.6 for any choice of α and β except the trivial choice $\alpha = 0$ and is therefore physically irrelevant. It might have some relevance in the case of a non-perfect fluid. We will not investigate that possibility in the present work.

11-A.6 The Solution Expressed in the Original Radial Coordinate

The solution that we have obtained for the generalized Roberson-Walker case with radial coordinate ř can be related back to the generalized Robertson-Walker solution using the original radial coordinate r. Eq. 11-A.4.7 implies

$$b_0(r) = \check{r}\, b(\check{r})/r \qquad (11\text{-A.6.1})$$

$$= 2a\,[1 + a^2 + (1 - a^2)(1 - kr^2)^{1/2}]^{-1} \qquad (11\text{-A.6.2})$$

using eqns. 11-A.4.10 and 11-A.5.4, and defining the constant a as

$$a = \gamma/\delta \qquad (11\text{-A.6.3})$$

An important special case of eq. 11-A.6.2 is the case where $a = 1$ (Ockham's Razor!). In this case we find eq. 11-A.6.2 becomes

$$b_0(r) = 1 \qquad (11\text{-A.6.4a})$$

and

$$A_0(t, r) = a(t) \qquad (11\text{-A.6.4b})$$

thus *recovering the normal Robertson-Walker solution exactly as a special case from eqns. 11-A.2.2, 11-A.4.6, and 11-A.5.2*. In this case we see

$$b(\check{r}) = 2k^{-1/2}(1 + \check{r}^2)^{-1} \qquad (11\text{-A.6.5})$$

from eq. 11-A.5.4 in the ř, θ, ϕ coordinate system. We considered this case within the expanded framework of a quantized model of the beginning of the universe in chapter 19 where it has non-trivial consequences.

11-A.7 Equivalence of the General Solution with the Original Robertson-Walker Solution

The general solution of the Einstein equations in the case of a perfect fluid (eqns. 11-A.4.13 – 11-A.4.17) for the scale factor $A(t, \check{r})$ in the generalized metric specified by eq. 11-A.4.9

$$d\tau^2 = dt^2 - A^2(t, \check{r})[d\check{r}^2 + \check{r}^2(d\theta^2 + \sin^2\theta\, d\varphi^2)] \qquad (11\text{-A.4.9})$$

is given by

$$A(t, \check{r}) = a(t)b(\check{r}) = 2\alpha k^{-1/2} a(t)[1 + \alpha^2\check{r}^2]^{-1} \qquad (11\text{-A.7.1})$$

where a is a constant given by eq. 11-A.6.3 and where a(t) is the solution of the standard equation (eq. 11-A.5.2) for the time dependent scale factor in the case of the Robertson-Walker metric. We now note that if we define a new radial vector

$$\text{ŗ} = \alpha\check{r} \qquad (11\text{-A.7.2})$$

then

$$d\tau^2 = dt^2 - a^2(t)4k^{-1}[1 + \text{ŗ}^2]^{-1}\,[d\text{ŗ}^2 + \text{ŗ}^2(d\theta^2 + \sin^2\theta\, d\varphi^2)] \qquad (11\text{-A.7.3})$$

Comparing this invariant interval expression with the $\alpha = 1$ expression for $b(\check{r})$ given in eq. 11-A.6.5 we conclude that we have proved the following theorem:

Theorem: The solution of the Einstein equations for the case of a perfect fluid for the generalized Robertson-Walker metric (eq. 11-A.2.2) is equivalent to a solution of the Einstein equations for the case of a perfect fluid in the case of the Robertson-Walker metric with a scale factor a(t) that is solely dependent on time.

In the case of *classical* gravitation theory the solutions are fully equivalent and related by a simple change of radial coordinates. Thus the homogeneity condition that we had relinquished at the beginning of our discussion is reinstated and the solution of the generalized case is consistent with a *maximally symmetric three-dimensional space*. The origin of the coordinate system can be chosen to be any point in space.

In the case of quantized versions of the Robertson-Walker model and our generalization of it we will see that the solutions are *generally not equivalent*. We explored a particular example of a quantized gravitational model that illustrates this point in chapter 11. The quantized model, which we defined, should be viewed as a first attempt to explore the quantum regime existing at the beginning of time – the Big Bang. The reasonableness of its results suggests that we are on the right track for understanding the Big Bang Epoch.

11-A.8 Hubble's Law in the Generalized Robertson-Walker Model

Our generalized Robertson-Walker metric assumes an inhomogeneous space with some fixed center at ř = 0. Presumably this center was the point at which the Big Bang took place at the beginning of the universe. In this section see how Hubble's Law emerges in the generalized model.

Hubble's law is one of the cornerstones of modern cosmology. While one might think a scale factor that depended on the radius coordinate might not be consistent with Hubble's Law it is easy to show that Hubble's Law is satisfied provided that the scale factor factorizes as required by the tr-component of Einstein's equations (eq. 11-A.4.14) for our generalized Robertson-Walker model.

First we give a simple derivation of Hubble's law for the case of a separable scale factor:

$$A(t, \check{r}) = a(t)b(\check{r}) \tag{11-A.4.12}$$

under the assumption that some remote galaxy lies on the same radial line as the line from the origin of the space coordinates to our galaxy. The proper distance between the remote galaxy and our galaxy has the form

$$D(t) = D_0 A(t, \check{r}) \tag{11-A.8.1}$$

The rate of recession of the remote galaxy is then

$$v = dD/dt = D_0 b(\check{r}) da(t)/dt = HD(t) \tag{11-A.8.2}$$

with

$$H = d \ln a(t)/dt \tag{11-A.8.3}$$

H is Hubble's constant. If the speed of recession is small then it determines the first-order Doppler shift. If we denote the wavelength of the received radiation as λ_r and the wavelength of the radiation at the source as λ_s then the shift z is

$$z = \lambda_r/\lambda_s - 1 = v/c = HD/c \tag{11-A.8.4}$$

Eq. 11-A.8.4 is Hubble's Law: the red shift of a galaxy is proportional to its distance. Note Hubble's Law in the generalized case follows from the factorization of A(t, ř) with a(t) satisfying the same differential equation as the Robertson-Walker scale factor.

Since a change of radial coordinate reduces the classical generalized model to the Robertson-Walker model, Hubble's Law can be proven in the general case of the non-collinearity of the coordinate origin, source and reception points.

12. Big Bang Scale Factor from t = 0 to the Present

12.1 Introduction

This chapter develops a numerical model for the scale factor from the Big Bang to the present. Although the universe that we live in is almost flat according to recent WMAP experimental data, the small curvature of space closes the universe and has significant effects. Therefore we will not use a flat space approximation. Our goal is to obtain an order of magnitude understanding of the evolution of the universe from the beginning. Our calculated numerical quantities appear generally to be of the right order of magnitude. And the physical ideas appear to be consistent with a reasonable view of reality.

Chapter 11 describes the physical implications of our blackbody Y quanta Dark Energy stabilization and expansion mechanisms for the universe. An especially important result of chapter 7 of Blaha (2004) for the expansion of the universe is: **Gravity is a repulsive force (anti-gravity!) at distances less than 9.08×10^{-34} cm.** (See Fig. 18.6.2 and Fig. 7.3.9.3 of Blaha (2004).) Thus the expansion of the universe gets an additional boost from gravity at ultra-short distances.

The data that we use in this chapter and throughout are the combined results of the WMAP and SDSS data[121] based on the assumption of a non-flat space. In particular we use the following values:

$$\begin{aligned} h &= \text{Hubble parameter} = 0.678 \\ \rho_{cr} &= \text{Critical density} = 1.8784h^2 \times 10^{-29}\ \text{g/cm}^{-3} \\ \Omega_\Lambda &= \text{Dark Energy density}/\rho_{cr} = \rho_{de}/\rho_{cr} = 0.692 \pm 0.012 \\ \Omega_d &= \text{Dark matter density}/\rho_{cr} = \rho_d/\rho_{cr} = 0.1186\ h^{-2} \\ \Omega_b &= \text{Baryon density}/\rho_{cr} = \rho_b/\rho_{cr} = 0.02226\ h^{-2} \\ \Omega_m &= \text{Matter density}/\rho_{cr} = \rho_m/\rho_{cr} = 0.308 \pm 0.012 \\ \Omega_{tot} &= \Omega_m + \Omega_\Lambda = 1.000 \pm 0.024 \\ t_0 &= t_{now} = \text{Age of universe} = 13.80 \pm 0.04\ \text{Gyr} \end{aligned} \tag{12.1.1}$$

The reader is directed to the original paper for other parameters, error bars and a detailed analysis of the data. *We use units where $\hbar = c = 1$ unless stated otherwise.*

---

[121] 2016 Particle Data Group, Particle Physics Booklet. Certain parameters have changed in the past 13 years of experiments and observations. However the changes do not significantly change the discussions and conclusions of this chapter and the following chapters.

In addition to the above input values we will use:[121,122]

$$\Omega_\gamma = \text{radiation density}/\rho_{cr} = \rho_\gamma/\rho_{cr} = 2.473h^{-2} \times 10^{-5} (T/2.7255)^4 h^{-2}$$
$$= 5.38 \times 10^{-5} \qquad (12.1.2)$$

and, also, based on an analysis of WMAP[123] data

$$r_{universe}(t_{now}) = \text{current radius of the universe} > 7.4 \times 10^{28} \text{ cm} \qquad (12.1.3)$$

12.2 The Behavior of the Scale Factor a(t) after the Big Bang Period

The universe, as we know it today, contains a variety of forms of energy. The current densities of these forms of energy are listed in section 12.1. From them we can develop the form of the total energy density and project it back to the instants after the Big Bang. The Big Bang period is significantly different as we have seen in the preceding chapter.

The total energy density as a function of time is

$$\rho_{tot} = [\Omega_\gamma/a^4(t) + \Omega_m/a^3(t) + \Omega_\Lambda]\rho_{cr} \qquad (12.2.1)$$

based on well-known arguments.[124] The time dependence of the scale parameter is given by the Einstein equation:

$$\dot{a}^2 - 8\pi G\rho_{tot}a^2/3 = -k \qquad (11\text{-A.5.2})$$

where a(t) is the Robertson-Walker scale factor with $a(t_{now}) = 1$.

Before proceeding to the solution of eq. 9-A.5.2 we need to obtain a reasonable estimate of the curvature constant k. We can use the Robertson-Walker expression for the radius of the universe

$$r_{universe}(t) = a(t)/k^{1/2} \qquad (12.2.2)$$

evaluated for the present time where eq. 12.1.3 sets a lower bound on the radius to obtain

$$k^{-1/2} > 7.4 \times 10^{28} \text{ cm} \qquad (12.2.3)$$

If we *assume* the actual radius of the universe is twice the lower bound, 1.48×10^{29} cm, we get

---

[122] See also Dodelson(2003) p. 41.
[123] N. J. Cornish, D. N. Spergel, G. D. Starkman, and E. Komatsu, Phys. Rev. Lett. **92**, 201302-1 (2004).
[124] Weinberg(1972), Dodelson(2003).

$$k = (1.48 \times 10^{29})^{-2}\ \text{cm}^{-2} = 4.57 \times 10^{-59}\ \text{cm}^{-2} \qquad (12.2.4)$$

We will use this value of k in the following sections.

12.2.1 General Form of a(t) Scale Factor Einstein Equation

We find the general form of the scale factor differential equation by combining eqns. 11-A.5.2 and 12.2.1

$$\dot{a}^2 - H_0^2 a^2(t)[\Omega_\gamma/a^4(t) + \Omega_m/a^3(t) + \Omega_\Lambda] = -k \qquad (12.2.1.1)$$

where Hubble's constant H_0 satisfies

$$H_0^2 = [d(\ln a)/dt]|_{t\,=\,t_{now}} = 8\pi G\rho_{cr}/3 \equiv 1.2 \times 10^{-56} h^2\ \text{cm}^{-2} \qquad (12.2.1.2a)$$

or

$$H_0 = 1.1 \times 10^{-28} h\ \text{cm}^{-1} \equiv 3.2 \times 10^{-18} h\ \text{s}^{-1} \qquad (12.2.1.2b)$$

If we evaluate eq. 12.2.1.1 for the present time we find

$$k = [\Omega_\gamma + \Omega_m + \Omega_\Lambda - 1]H_0^2 \cong [\Omega_m + \Omega_\Lambda - 1]H_0^2 = (0.01)H_0^2 \,{}^{+0.087}_{-0.082}$$

$$= 6.1 \times 10^{-59}\ \text{cm}^{-2} \,{}^{+44.2}_{-41.7} \qquad (12.2.1.2c)$$

Notice the error "bars" make the value of k uncertain. They result from the difficulties in experimentally measuring the densities Ω_m and Ω_Λ.

The range of values for k in eq. 12.2.1.2c is $[-47.8 \times 10^{-59}, 50.3 \times 10^{-59}]$. Therefore we feel that the radius value determined from WMAP data in eq. 12.1.3, even though it is a lower bound, may be a better indication of the value of k assuming a closed Robertson-Walker universe. In any case we will use this value for k in eq. 12.2.4. This value is only 25% different from the estimate in eq. 12.2.1.2c and thus well within the order of magnitude goal of our calculations. The value of k that we have selected:

$$k = 4.6 \times 10^{-59}\ \text{cm}^{-2} \qquad (12.2.4)$$

implies

$$\Omega_m + \Omega_\Lambda = 1.00$$

which is within the error bars of these quantities.

We now define

$$\xi = k/H_0^2 = 3.9 \times 10^{-3} h^{-2} \qquad (12.2.1.3)$$

for later use.

The solution of eq. 12.2.1.1 can be put into the form of integrals representing combinations of elliptic integrals:

$$\int_{a(t')}^{a(t)} da\ a\ H_0^{-1}\ [\Omega_\Lambda a^4 - \xi a^2 + \Omega_m a + \Omega_\gamma]^{-1/2} = \int_{t'}^{t} dt \qquad (12.2.1.4)$$

The result of these integrations is an implicit equation for a(t) that cannot be expressed in a simple closed form in terms of known functions. This equation can be easily solved numerically. A graph of a(t) is displayed in Fig. 12.2.1.1. Although it looks linear there are significant non-linearities in various parts of the plot of a(t). The radiation-dominated phase is not visible. It is a small slice of the plot since it amounts to less than 10^{13} s.

Because we know physically that approximations are possible for each of the various epochs: the matter-dominated epoch, the radiation-dominated epoch and so on, we can find physically meaningful approximations for each epoch. We therefore provisionally divide the life of the universe into two epochs: an explosive growth epoch, and an expanding epoch subdivided into matter-dominated and radiation-dominated phases. We would have subdivided the expanding epoch into three phases if the constant ξ were not so small.

12.2.2 The Explosive Growth Epoch

The integral on the left in eq. 12.2.1.4 appears to support a simple approximation during the explosive growth phase. Notice that at $t = t_{now}$ we have $-\xi a^2 + \Omega_m a + \Omega_\gamma \cong .308$. The sum of these terms gets smaller as we proceed into the past. Therefore we approximate the left side with

$$\int_{a(t')}^{a(t)} da\ aH_0^{-1}[\Omega_\Lambda a^4]^{-1/2} = (H_0\Omega_\Lambda^{1/2})^{-1} \int_{a(t')}^{a(t)} da/a = (H_0\Omega_\Lambda^{1/2})^{-1}\ln[a(t)/a(t')] \cong t - t' \qquad (12.2.2.1)$$

or

$$a(t) = a(t')\exp[H_0\Omega_\Lambda^{1/2}(t - t')] \qquad (12.2.2.2)$$

Thus we have deSitter-like exponential growth. We would expect this growth phase to last approximately for the period where

$$\Omega_\Lambda a^4(t_E) \approx \Omega_m a(t_E) \qquad \Rightarrow \qquad a(t_E) = .77 \qquad (12.2.2.3)$$

until the present where t_E is the time when $a(t_E) = .77$. Since we have normalized a(t) by

$$a(t_{now}) = 1 \qquad (12.2.2.4)$$

we see that eq. 12.2.2.2 requires

$$a(t) = \exp[H_0\Omega_\Lambda^{1/2}(t - t_{now})] \tag{12.2.2.5}$$

in the epoch that we have called the Explosive Growth Epoch. Note

$$H_0\Omega_\Lambda^{1/2} = 1.8 \times 10^{-18}\ s^{-1} \tag{12.2.2.6}$$

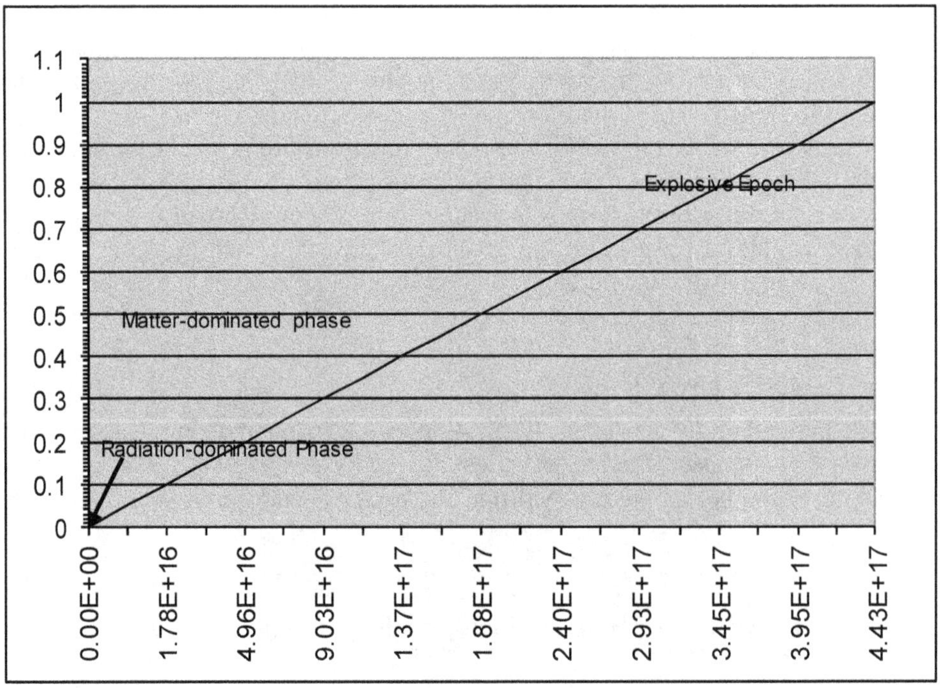

Figure 12.2.1.1. A plot of a(t) generated from eq. 12.2.1.4 through numerical integration.

Epoch	Type	Phases
I	**Explosive Growth**	
II	**Expanding**	**Matter-dominated Radiation-dominated**

Table 12.2.1.1. Epochs and phases of the universe after the Big Bang Epoch.

The beginning of this epoch is set by eq. 12.2.2.3:

$$t_E = t_{now} + (H_0\Omega_\Lambda^{1/2})^{-1}\ln(.77) = 2.99 \times 10^{17} \text{ s} \qquad (12.2.2.7)$$

giving the time interval of this epoch as 1.47×10^{17} s = 4.7 Gyr —much longer than the radiation-dominated era, and still expanding. The transition time t_E is close to the standard hypothesis for the appearance of Dark Energy of at a red-shift z ~ .5 or a(t) = 2/3. ($1 + z = a^{-1}$) The a(t) value of 2/3 corresponds to a time $t_E = 2.7 \times 10^{17}$ s by eq. 12.2.1.4 (numerical solution) which is to be compared to our estimate $t_E = 2.99 \times 10^{17}$ s – a roughly 10% difference.

12.2.3 The Expanding Epoch

The Expanding Epoch includes the matter-dominated and radiation-dominated phases. The solution for the scale factor in this epoch is obtained by neglecting the Ω_Λ term in eq. 12.2.1.4:

$$\int_{a(t')}^{a(t)} da\, a\, H_0^{-1} \left[-\xi a^2 + \Omega_m a + \Omega_\gamma\right]^{-1/2} = \int_{t'}^{t} dt \qquad (12.2.3.1)$$

This equation is easily integrated yielding:

$$[-\xi a^2(t) + \Omega_m a(t) + \Omega_\gamma]^{1/2} + \Omega_m(2\xi^{1/2})^{-1}\arcsin[(\Omega_m - 2\xi a(t))(\Omega_m^{\ 2} + 4\xi\Omega_\gamma)^{-1/2}] -$$

$$- \ [-\xi a^2(t') + \Omega_m a(t') + \Omega_\gamma]^{1/2} - \Omega_m(2\xi^{1/2})^{-1}\arcsin[(\Omega_m - 2\xi a(t'))(\Omega_m^{\ 2} + 4\xi\Omega_\gamma)^{-1/2}] =$$

$$= -\xi H_0(t - t') \qquad (12.2.3.2)$$

Assuming a(0) = 0 and letting $t_0 = 0$ we find

$$[-\xi a^2(t) + \Omega_m a(t) + \Omega_\gamma]^{1/2} + \Omega_m(2\xi^{1/2})^{-1}\arcsin[(\Omega_m - 2\xi a(t))(\Omega_m^{\ 2} + 4\xi\Omega_\gamma)^{-1/2}] -$$

$$- \Omega_\gamma^{1/2} - \Omega_m(2\xi^{1/2})^{-1}\arcsin[\Omega_m(\Omega_m^{\ 2} + 4\xi\Omega_\gamma)^{-1/2}] =$$

$$= -\xi H_0 t$$

$$(12.2.3.3)$$

Eq. 12.2.3.3 can be substantially simplified. Note the arguments of the arcsines are both near one in value due to the smallness of Ω_γ and ξ. Both arcsines can be approximated using

$$\arcsin(1 - \epsilon) \cong \pi/2 - (2\epsilon)^{1/2} \qquad (12.2.3.4)$$

for small ϵ.

First we approximate the arguments of the arcsines with

$$\arcsin[(\Omega_m - 2\xi a(t))(\Omega_m^{\ 2} + 4\xi\Omega_\gamma)^{-1/2}] \cong \arcsin(1 - 2\xi a(t)\Omega_m^{\ -1} - 2\xi\Omega_\gamma\Omega_m^{\ -2})$$

and

$$\arcsin[\Omega_m(\Omega_m^{\ 2} + 4\xi\Omega_\gamma)^{-1/2}] \cong \arcsin(1 - 2\xi\Omega_\gamma\Omega_m^{\ -2})$$

Then using eq. 12.2.3.4 we obtain

$$[-\xi a^2(t) + \Omega_m a(t) + \Omega_\gamma]^{1/2} - [\Omega_m a(t) + \Omega_\gamma]^{1/2} \cong -\xi H_0 t \qquad (12.2.3.5)$$

Noting that $\xi a^2(t)$ is much smaller than the other terms in the first square root in eq. 12.2.3.5 we can further approximate that equation by expanding the square root to obtain:

$$a^2(t)[\Omega_m a(t) + \Omega_\gamma]^{-1/2} \cong 2H_0 t \qquad (12.2.3.6)$$

Eq. 12.2.3.6 embodies the standard matter-dominated and radiation-dominated expressions for the scale factor.

In the case of the matter-dominated phase we have

$$\Omega_m a(t) > \Omega_\gamma$$

and can approximate eq. 12.2.3.6 accordingly

$$a^2(t)[\Omega_m a(t)]^{-1/2} \cong 2H_0 t \qquad (12.2.3.7)$$

or

Matter-dominated Phase

$$a(t) \cong [2\Omega_m^{\ 1/2} H_0 t]^{2/3} \qquad (12.2.3.8)$$

In the case of the radiation-dominated phase we have

$$\Omega_m a(t) < \Omega_\gamma$$

due to the smallness of the scale factor in that phase We approximate eq. 12.2.3.6 accordingly

$$a^2(t)[\Omega_\gamma]^{-1/2} \cong 2H_0 t \qquad (12.2.3.9)$$

Thus

Radiation-dominated Phase

$$a(t) \cong [2\Omega_\gamma^{\ 1/2} H_0 t]^{1/2} \qquad (12.2.3.10)$$

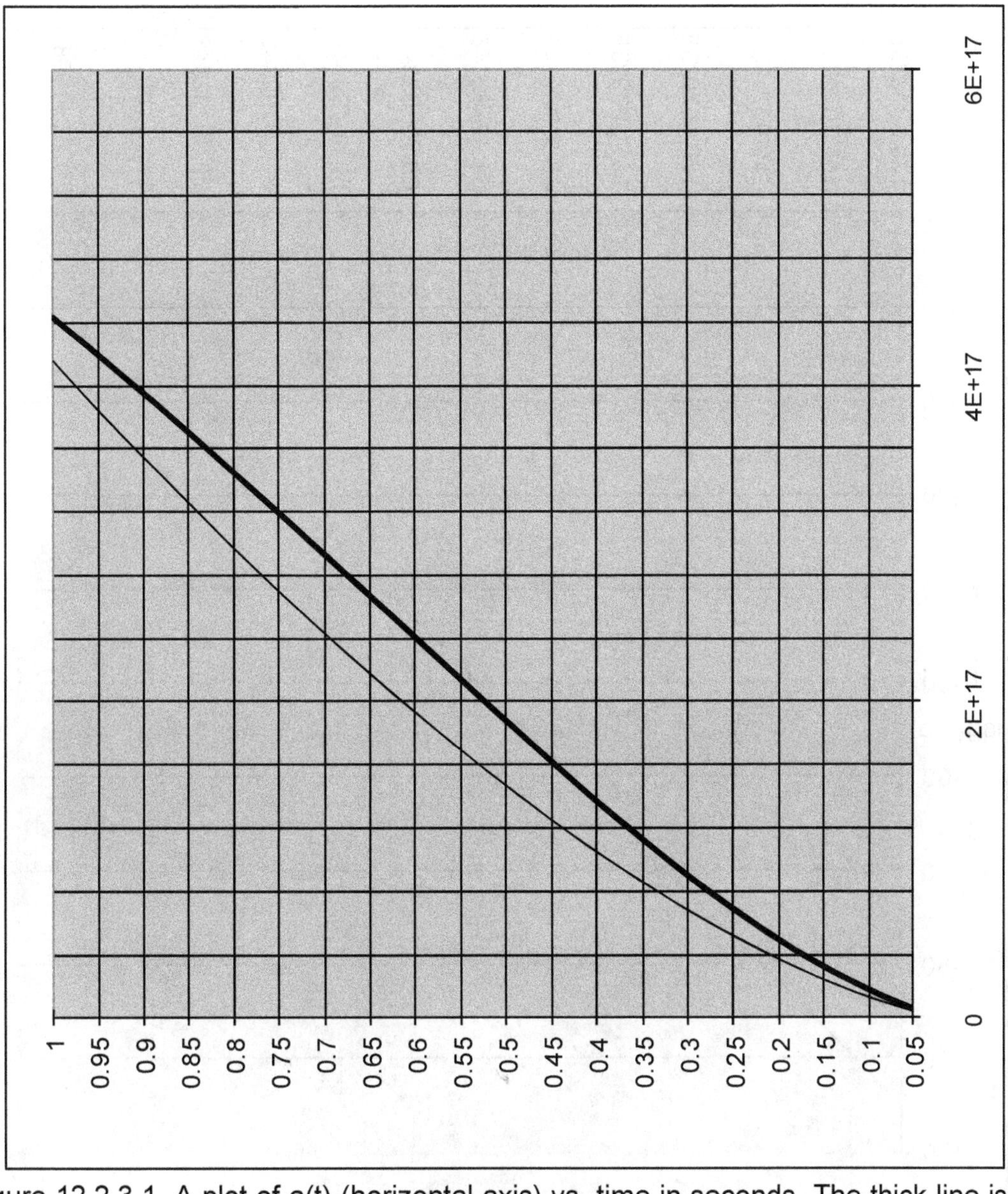

Figure 12.2.3.1. A plot of a(t) (horizontal axis) vs. time in seconds. The thick line is the plot of a(t) obtained by direct numerical integration of eq. 12.2.1.4 including the three density terms and the curvature constant term. The thin line is a plot of a(t) calculated directly from the approximation eq. 12.2.3.6. The approximation becomes increasingly better for small times as $t \to 0$. (Reader: please rotate page 90 degrees clockwise.)

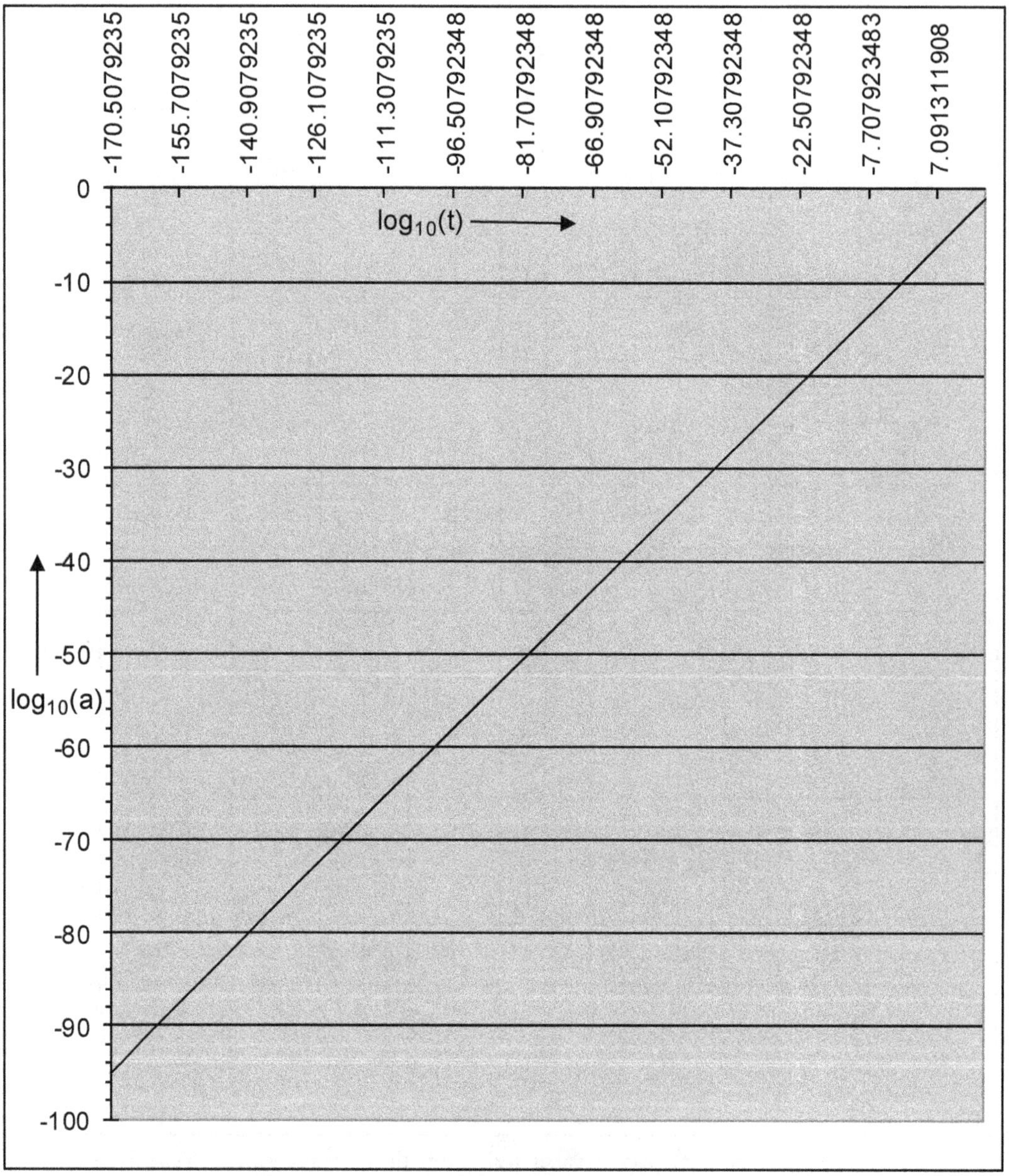

Figure 12.2.3.2. A log-log (base 10) plot of a(t) vs. t (in seconds) for small times calculated from eq. 12.2.3.6.

The crossover point between the radiation-dominated and matter-dominated phase is at

$$\Omega_m a(t_{RM}) = \Omega_\gamma \quad \text{or} \quad a(t_{RM}) = 1.8 \times 10^{-4} \qquad (12.2.3.11)$$

where t_{RM} is the crossover time:

$$t_{RM} = 7 \times 10^{11}\ \text{s} \qquad (12.2.3.12)$$

Figs. 12.2.3.1 and 12.2.3.2 contain plots of a(t). Fig. 12.2.3.1 shows the approximate implicit equation for a(t) (eq. 12.2.3.6) is quite good over the entire range, and particularly good for small times in the radiation-dominated time frame. Since this region is the region of interest as it connects to the Big Bang Epoch we shall use this approximation, and eq. 12.2.3.10, for a(t) at very small times near t = 0.

12.3 Two Tier Quantum Big Bang Model at the Beginning of Time

Section 12.2 presented the approximate expressions for the scale factor near t = 0. In this section we will consider numerical estimates for the scale factor, and other quantities of interest, such as the temperature and density near, and at, t = 0 in the Two Tier model whose early time behavior is described in chapter 11.

In estimating quantities we are confronted with an imprecise determination of the needed input parameters because of experimental uncertainties and the impossibility of currently finding certain parameters experimentally in a model independent way. So we will use reasonable estimates for input parameters realizing that they may sometimes be off by up to a few orders of magnitude. Because of the vast differences in value between terms comprising the scale factors we will see a few orders of magnitude is often not a significant issue in determining the relative importance of terms.

The reader may notice slight differences in values due to rounding off numbers to three figures in the text while keeping values to 16 significant digits in the calculations. These differences can have a cumulative effect so we want to emphasize that our goal is order of magnitude accuracy.

The input data items are those of section 12.2 plus:

1. The current temperature of the Cosmic Microwave Background (CMB) radiation = 2.725 °K.

2. We assume $M_c = M_{Planck} = 1.22 \times 10^{28}$ ev since M_{Planck} is the only large mass intrinsic to the theory of gravitation and thus seemed to be a natural choice.

12.3.1 The CMB Temperature

From the current CMB temperature (T = 2.725 °K) we find

$$\kappa T = \kappa T_0/a(t_{now}) = \kappa T_0 = 2.3 \times 10^{-4} \text{ ev} \tag{12.3.1.1}$$

For later use we define

$$\varkappa = \pi^{3/2}\kappa T_0/(2M_c) \cong 5.3 \times 10^{-32} \tag{12.3.1.2}$$

12.3.2 The Generalized Robertson-Walker Scale Factor

A_{BB} is related to B_{BB} by eq. 11.86. Our approximation for the B_{BB} scale factor is:[125]

$$B_{BB}(t, y) \cong (1 + M_c^2y^2)^{-1}\{-i\varkappa M_cy + k^{-1/2}M_ca(t) + [\varkappa^2 - 2i\varkappa yk^{-1/2}M_c^2a(t) + k^{-1}M_c^2a^2(t)]^{1/2}\} \tag{12.1.5}$$

If we let $y = M_c^{-1}$ then we can find B_{BB} "at the borders of the universe" where it simplifies to:

$$B_{BB}(t) = B_{BB}(t, M_c^{-1}) \cong \{-i\varkappa + \varpi a(t) + [\varkappa^2 - 2i\varkappa\varpi a(t) + \varpi^2a^2(t)]^{1/2}\}/2 \tag{12.3.2.1}$$

with

$$\varpi = k^{-1/2}M_c = 9.2 \times 10^{61} \tag{12.3.2.2}$$

If t and a(t) are small,

$$A_{BB} = A_{BB}(t, \check{r}) = M_c^{-1}B_{BB}(t, \check{r}) \cong \beta_0(\check{r}) + \beta_1(\check{r})a(t) + \ldots \tag{12.3.2.3}$$

$$\beta_0(\check{r}) = \varkappa(1 - i\check{r})/[M_c(1 + \check{r}^2)] \tag{12.3.2.4}$$

$$\beta_1(\check{r}) = k^{-1/2}(1 - i\check{r})/(1 + \check{r}^2) \tag{12.3.2.5}$$

We noted earlier that if we transformed our results back to Robertson-Walker coordinates we would have a modified scale factor $a = a_{BBRW}$ which we can continue to express in terms of $\check{r} = M_cy$.

$$a(t) \rightarrow (1 + M_c^2y^2)k^{1/2}A_{BB}(t, y)/2 \equiv a_{BBRW}(t, \check{r}) \tag{11.5.2.4}$$

Thus

$$a_{BBRW}(t, \check{r}) = (1 + M_c^2y^2)(2k^{-1/2}M_c)^{-1}B_{BB}(t, y) \tag{12.3.2.6}$$

---

[125] Please note that the physical value of scale factors is the absolute value of the complex scale factors (obtained by use of Reality group transformations) appearing here and in the following discussions.

$$= \tfrac{1}{2}\{a(t) - i\chi\varpi^{-1}\check{r} + [(\chi/\varpi)^2 + a^2(t) - 2i(\chi/\varpi)\check{r}a(t)]^{1/2}\} \quad (12.3.2.7)$$

by eq. 11.92. If we evaluate a_{BBRW} at $\check{r} = M_c y = 1$ (the maximum value of y), since it determines the "size" of the universe, then

$$a_{BBRW}(t) \equiv a_{BBRW}(t, 1) = \{a(t) - i\gamma + [\gamma^2 + a^2(t) - 2i\gamma a(t)]^{1/2}\}/2 \quad (12.3.2.8)$$

with the dimensionless constant

$$\gamma = \chi/\varpi = 5.8 \times 10^{-94} \quad (12.3.2.9)$$

This value reminds one of Eddington's famous remark that cosmological quantities often have orders of magnitude that are approximately multiples of 90.

The real and imaginary parts of $a_{BBRW}(t)$ are:

$$\text{Re } a_{BBRW}(t) = a(t)/2 + [R(t)(1 + \cos\psi(t))/2]^{1/2}/2 \quad (12.3.2.10)$$

and

$$\text{Im } a_{BBRW}(t) = -\gamma/2 - [R(t)(1 - \cos\psi(t))/2]^{1/2}/2 \quad (12.3.2.11)$$

where

$$R(t) = [(\gamma^2 + a^2(t))^2 + 4\gamma^2 a^2(t)]^{1/2} \quad (12.3.2.12)$$

and

$$\cos\psi(t) = (\gamma^2 + a^2(t))/R \quad (12.3.2.13)$$

There are 2 distinctly different periods specified by the time dependence of a_{BBRW}. The first period corresponds to a universe of "slowly" increasing size. The second period is the radiation and matter dominated phases with a fairly rapid increase in a(t). The boundary time t_c between these periods is specified by:

$$\gamma = a(t_c) \quad (12.3.2.14)$$

yielding

$$t_c = 1.15 \times 10^{-167}\ \text{s} \cong 10^{-167}\ \text{s} \quad (12.3.2.15)$$

In this period we find

$$\text{Re } a_{BBRW}(0) \cong \gamma/2 = 2.9 \times 10^{-94} \quad (12.3.2.16a)$$

$$\text{Re } a_{BBRW}(t_c) \cong 1.28\gamma = 7.5 \times 10^{-94} \quad (12.3.2.16b)$$

$$\text{Im } a_{BBRW}(0) = -\gamma/2 = -2.9 \times 10^{-94} \quad (12.3.2.16c)$$

with the radius, and volume, of the universe "slowly" increasing during this period. The nature of this period directly reflects the effects of the blackbody Y quanta. This can be seen from its inverse square dependence on M_c. As $M_c \to \infty$ the constant $\gamma \to 0$ and thus $a_{BBRW} \to 0$ with the universe scaling down to zero size with the attendant catastrophes of infinite density and temperature that appear in the standard models. The blackbody quanta (Dark Energy) give the universe a meta-stable initial size thus avoiding catastrophic divergences.

Epoch	Type	Phases	Time Period
I	Explosi ve Growth	Dark Energy-dominated	3×10^{17} s – 4.5×10^{17} s
I I	Expand ing	Matter-dominated Radiation-dominated	7×10^{11} s – 3×10^{17} s 1.1×10^{-167} s – 7×10^{11} s
III	Metastable Big Bang	Blackbody Y quanta (Dark Energy) dominated	0 s – 1.1×10^{-167} s

Table 12.3.2.1 Epochs and phases of the Universe since t = 0.

The period after t_c is dominated by the usual scale factor a(t). This scale factor shows up directly in the real part of a_{BBRW}. Its behavior is:

$t < t_c$ (or $a(t) < \gamma$)

$$\text{Re } a_{BBRW}(t) \cong \gamma/2 + a(t)/2 \qquad (12.3.2.17a)$$

$t_c < t < t_{now}$

$$\text{Re } a_{BBRW}(t) \cong a(t) \qquad (12.3.2.17b)$$

The behavior of the imaginary part of a_{BBRW} is also indirectly dominated by a(t) but in a much less dramatic way. The gradual growth of the imaginary part of a_{BBRW} is more or less indicated by the following three values:

$$\text{Im } a_{BBRW}(0) \cong -\gamma/2 = -2.9 \times 10^{-94} \qquad (12.3.2.16c)$$
$$\text{Im } a_{BBRW}(t_c) \cong -0.822\gamma = -4.8 \times 10^{-94} \qquad (12.3.2.18a)$$
$$\lim_{t \to \infty} \text{Im } a_{BBRW}(t) = -\gamma = -5.8 \times 10^{-94} \qquad (12.3.2.18b)$$

and also the behavior

$a(t) \ll \gamma$ (or $t < t_c$)

$$\text{Im } a_{BBRW}(t) \cong -\gamma/2 - a(t)/2 \quad (12.3.2.18c)$$

$\gamma \ll a(t)$ (or $t_c < t$)

$$\text{Im } a_{BBRW}(t) \cong -\gamma + O([\gamma/a(t)]^2) \cong -\gamma \quad (12.3.2.18d)$$

Both the real and imaginary parts of a_{BBRW} roughly double in the time period $[0, t_c]$ and thereafter we see a gradual increase of Im a_{BBRW} in absolute value from $-.5\gamma$ to $-\gamma$ over the lifetime of the universe.

After t_c the real part grows dramatically (eq. 12.3.2.17) while the imaginary part remains minute. The details of the interpretation of the behavior of the scale factor a_{BBRW} in the Big Bang Epoch $[0, t_c]$ will be explored in chapter 13. It suffices, for now, to say the universe in the period before t_c is in a meta-stable state of "slowly" growing size due to the dynamics of the blackbody Y quanta. At t_c the epoch of the expanding universe as we know it begins!

In differentiating between the real and imaginary parts of a_{BBRW} we must realize that the Reality group combines them into a single real quantity when scaling coordinates. Happily the small size of the imaginary part it can be neglected in most situations.

Before proceeding to describe physically interesting features of the early universe we note the radial dependence of the scale factor $a_{BBRW}(t, \check{r})$ in general:

$$a_{BBRW}(t, \check{r}) = \{-i\gamma\check{r} + a(t) + [\gamma^2 - 2i\gamma\check{r}a(t) + a^2(t)]^{1/2}\}/2 \quad (12.3.2.19)$$

$$\text{Re } a_{BBRW}(t, \check{r}) = a(t)/2 + [R(t, \check{r})(1 + \cos\psi(t, \check{r})/2]^{1/2}/2 \quad (12.3.2.20)$$

$$\text{Im } a_{BBRW}(t, \check{r}) = -\gamma\check{r}/2 - [R(t, \check{r})(1 - \cos\psi(t, \check{r}))/2]^{1/2}/2 \quad (12.3.2.21)$$

where

$$R(t, \check{r}) = [(\gamma^2 + a^2(t))^2 + 4(\gamma\check{r}a(t))^2]^{1/2} \quad (12.3.2.22)$$

and

$$\cos\psi(t, \check{r}) = (\gamma^2 + a^2(t))/R(t, \check{r}) \quad (12.3.2.23)$$

We find that the scale factor $a_{BBRW}(t, \check{r})$ is approximated in various time periods to well within an order of magnitude by

$0 \le t < t_c$

$$\text{Re } a_{BBRW}(t, \check{r}) \cong \gamma/2 + a(t)/2 \quad (12.3.2.24)$$

$$\text{Im } a_{BBRW}(t, \check{r}) \cong -\gamma\check{r}/2 - a(t)\check{r}/2 \quad (12.3.2.25)$$

$t = t_c$

$$\text{Re } a_{BBRW}(t_c, \check{r}) = \{1 + [(1 + \check{r}^2)^{1/2} + 1]^{1/2}\}\gamma/2 \le 1.28\gamma = 7.5 \times 10^{-94}$$

$$0 \ge \text{Im } a_{BBRW}(t_c, \check{r}) = \{-\check{r} - [(1 + \check{r}^2)^{1/2} - 1]^{1/2}\}\gamma/2 \ge -0.8\gamma$$

$t_c < t$

$$\text{Re } a_{BBRW}(t, \check{r}) \cong a(t)\{1 + (1 + \check{r}^2)\gamma^2/4\} \cong a(t) \quad (12.3.2.26)$$

$$\text{Im } a_{BBRW}(t, ř) \cong -\gamma ř/2 - [ř^2 + \gamma^2/(4a^2(t))]^{½}\gamma/2 \qquad (12.3.2.27)$$

t → ∞

$$\text{Re } a_{BBRW}(t, ř) \cong a(t)\{1 + (1 + ř^2)\gamma^2/4\} \cong a(t) \qquad (12.3.2.28)$$

$$\text{Im } a_{BBRW}(t, ř) \cong -\gamma ř = -5.8 \times 10^{-94} ř \qquad (12.3.2.29)$$

12.3.3 Temperature of the Early Universe in the Generalized Robertson-Walker Metric

We will now examine the temperature of the early universe based on the results of the preceding section. For t near t = 0, we note κT depends on the blackbody scale factor of the generalized Robertson-Walker metric:[126]

$$\kappa T = \kappa T_0/B_{BB}(t, y) \qquad (12.3.3.1)$$

and

$$B_{BB}(t, y) = 2k^{-½}M_c(1 + M_c^2 y^2)^{-1} a_{BBRW}(t, ř) \qquad (12.3.3.2)$$

by eq. 12.3.2.6 using the variable

$$ř \equiv M_c y \qquad (12.3.3.3)$$

for convenience. Therefore from eqns. 12.3.2.24 – 12.3.2.27 we see

$0 \le t < t_c$

$$\kappa T_< \equiv \kappa T(0 \le t < t_c) \cong 2\kappa T_0(1 + iř)/\chi$$
$$= 8.8 \times 10^{27}(1 + iř) \text{ ev} \qquad (12.3.3.4)$$

$t_c < t$

$$\kappa T_> \equiv \kappa T(t_c < t)$$
$$\cong \kappa T_0(a(t) + i\gamma ř)/[2k^{-½}M_c(1 + ř^2)^{-1}(a^2(t) + \gamma^2 ř^2)]$$
$$\cong \kappa T_0(1 + ř^2)/[2k^{-½}M_c a(t)] = 1.3 \times 10^{-66}(1 + ř^2)/a(t) \text{ ev} \qquad (12.3.3.5)$$

We note that the temperatures calculated in this section are in generalized Robertson-Walker coordinates (eq. 11.2.4.5) and not in Robertson-Walker coordinates.

Note also that the physical temperature is the absolute value of the complex temperature due to the use of the Reality group.

---

[126] Here again we remind the reader that complex values for radii, temperature and so on are transformed by the Reality group to real values – their absolute values.

12.3.4 Consistency of Y_{BB} Approximation with the Resulting Temperature near t = 0

We now address the question of whether the values found for κT are consistent with the approximations made in chapter 11 in order to obtain an expression for Y_{BB}:

$$\cos(\omega\kappa Tt) \approx 1 \tag{11.66}$$
$$J_1(\omega\kappa Ty) \approx \omega\kappa Ty/2 \tag{11.67}$$

The first approximation is valid if $\kappa Tt \sim 0$ since the Planck distribution factor makes the largest contribution to the integral come from small ω. The values of $\kappa T_<$ and $\kappa T_>$ show that for larger times κTt is very small:

$$\kappa T_< t_c \leq 2 \times 10^{-124}$$
$$\kappa T_> t \leq (3.89 \times 10^{-51}\ s^{-1})t/a(t) \leq (3.9 \times 10^{-51}\ s^{-1})t_{now}/a(t_{now}) = 1.7 \times 10^{-33} \tag{12.3.4.1}$$

The second approximation is valid if $\kappa Ty \ll 1$ since the Planck distribution factor again makes the largest contribution to the integral come from small ω. We find the maximum values of κTy in the two time periods from eqns. 12.3.3.4 and 12.3.3.5 to be

$0 \leq t < t_c$

$$\text{MAX}(|\omega\kappa T_< y|) = |\omega\kappa T_< M_c^{-1}| = 1.0\omega \tag{12.3.4.2}$$

which is small since the dominant part of the integration comes from small ω; and

$t_c < t$

$$\text{MAX}(|\omega\kappa T_> y|) = |\omega\kappa T_>(t_c)M_c^{-1}| = 2.10 \times 10^{-94}\omega/a(t)$$
$$\leq 2.1 \times 10^{-94}\omega/a(t_c) = 0.4\omega \tag{12.3.4.3}$$

which is also small. Thus the Bessel function power series expansion is well approximated by its first term:

$$J_1(\omega\kappa Ty) \approx (\omega\kappa Ty/2) \tag{12.3.4.4}$$

We conclude our approximate calculation of A(t, y) is valid for all time and for the complete range of y values: $0 \leq y \leq M_c^{-1}$.

12.3.5 Plots of the Scale Factor from t = 0 to the Present

We will create several plots of the scale factor from t = 0 to the present time $t = 4.46 \times 10^{17}$ s for the maximum value of $y = M_c^{-1}$, which corresponds to the maximum Robertson-Walker radius coordinate value $r = k^{-½}$. The approximation that we have developed for B_{BB} has been justified for times between the Big Bang and the present.

In Fig. 12.3.5.1 we show a log – log plot of the real and imaginary parts of $a_{BBRW}(t)$ vs. t using base 10 logarithms for $t \in [10^{-200}, 10^{20}]$ seconds. In Fig. 12.3.5.2 we plot the real and imaginary parts of a_{BBRW} vs. time from t = 0 to $t = 1.2 \times 10^{-246}$ s. In Fig. 12.3.5.3 we plot the real part of a_{BBRW}, and the Robertson-Walker scale factor a(t), vs. time in seconds in the period around 10^{-167} s. It shows the rapidity of the transition of Re a_{BBRW} from slowly rising to rapidly rising with a(t).

12.4 The Interpretation of the Complex Scale Factor

The interpretation of the complex scale factor $a_{BBRW}(t, \check{r})$ hinges on its role in the expression for the proper interval. The expression for the proper interval in the Robertson-Walker metric is:

$$d\tau^2 = dt^2 - R^2(t)[dr^2/(1 - kr^2) + r^2(d\theta^2 + \sin^2\theta d\varphi^2)]$$

It was generalized to

$$d\tau^2 = dt^2 - A^2(t, \check{r})[d\check{r}^2 + \check{r}^2(d\theta^2 + \sin^2\theta\, d\varphi^2)] \qquad (11\text{-A.4.9})$$

and an identification was made between the Robertson-Walker scale factor $R(t) \equiv a(t)$ and $a_{BBRW}(t, \check{r})$ through the following chain of equalities and correspondences

$$a(t) = a(t)b_0(r) = A(t, \check{r})/b(\check{r}) \rightarrow A_{BB}(t, \check{r})(1 + \check{r}^2)k^{½}/2 = a_{BBRW}(t, \check{r}) \qquad (12.4.1)$$

where the → relates the classical expressions on the left with the expressions on the right that embody quantum corrections due to Y-quanta blackbody radiation. Eq. 12.4.1 is based on eqns. 11-A.4.6, 11-A.4.7, 11-A.4.9, 11-A.4.11, 11-A.4.12, and 11-A.6.4a. Combining eqns. 12.4.1 we find

$$d\tau^2 = dt^2 - a_{BBRW}(t, \check{r}(r))^2[dr^2/(1 - kr^2) + r^2(d\theta^2 + \sin^2\theta d\varphi^2)] \qquad (12.4.2)$$

where the relation between ř and r, ř(r), is specified by eq. 11-A.4.12. We call the metric in eq. 12.4.2 a generalized Robertson-Walker metric. It is equivalent to eq. 11-A.4.9 – differing only in the definition of the radial coordinate.

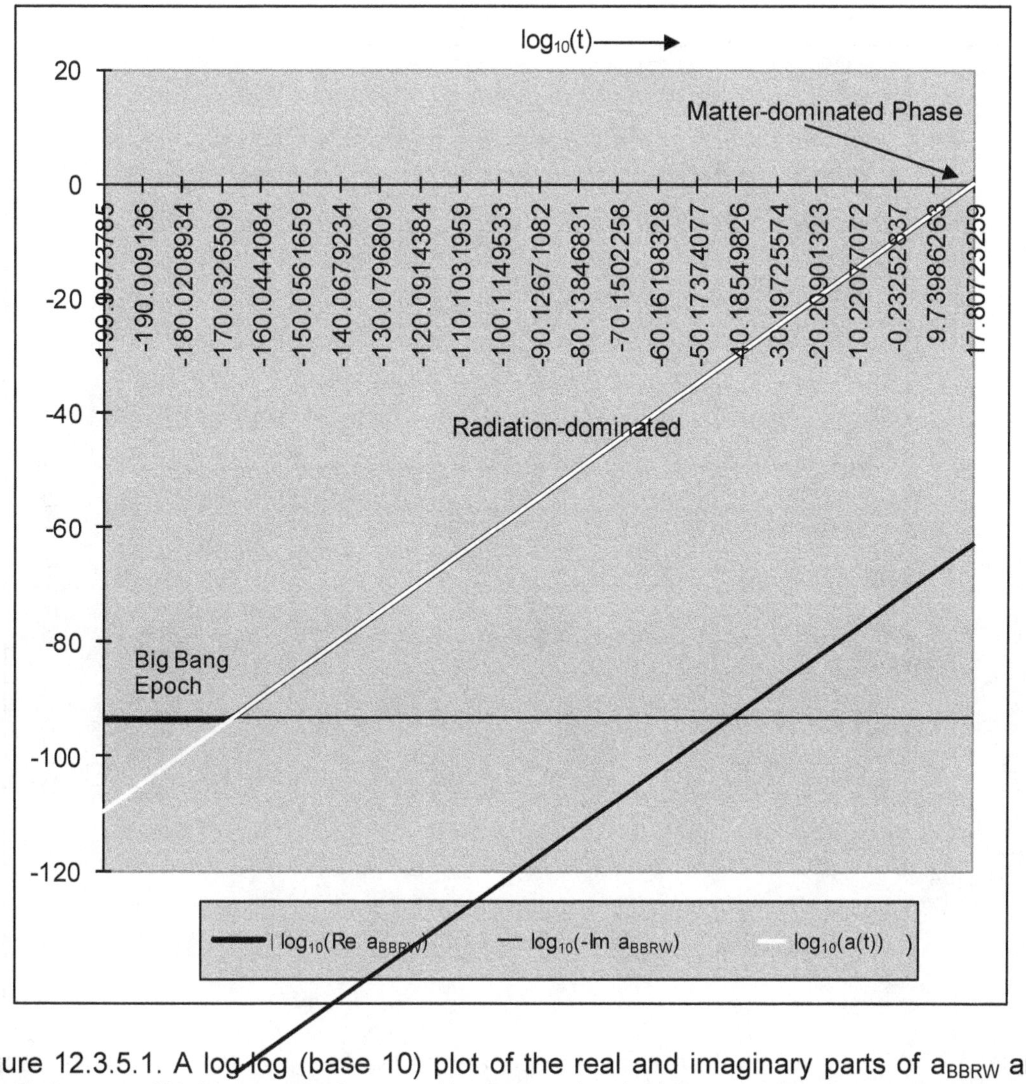

Figure 12.3.5.1. A log-log (base 10) plot of the real and imaginary parts of a_{BBRW} and the Robertson-Walker scale factor a(t) versus the log (base 10) of time in seconds. Note the imaginary part of a_{BBRW} is very slowly growing. The real part of a_{BBRW} is growing slowly until $t_c \approx 10^{-167}$ s and thereafter equals a(t) to good approximation.

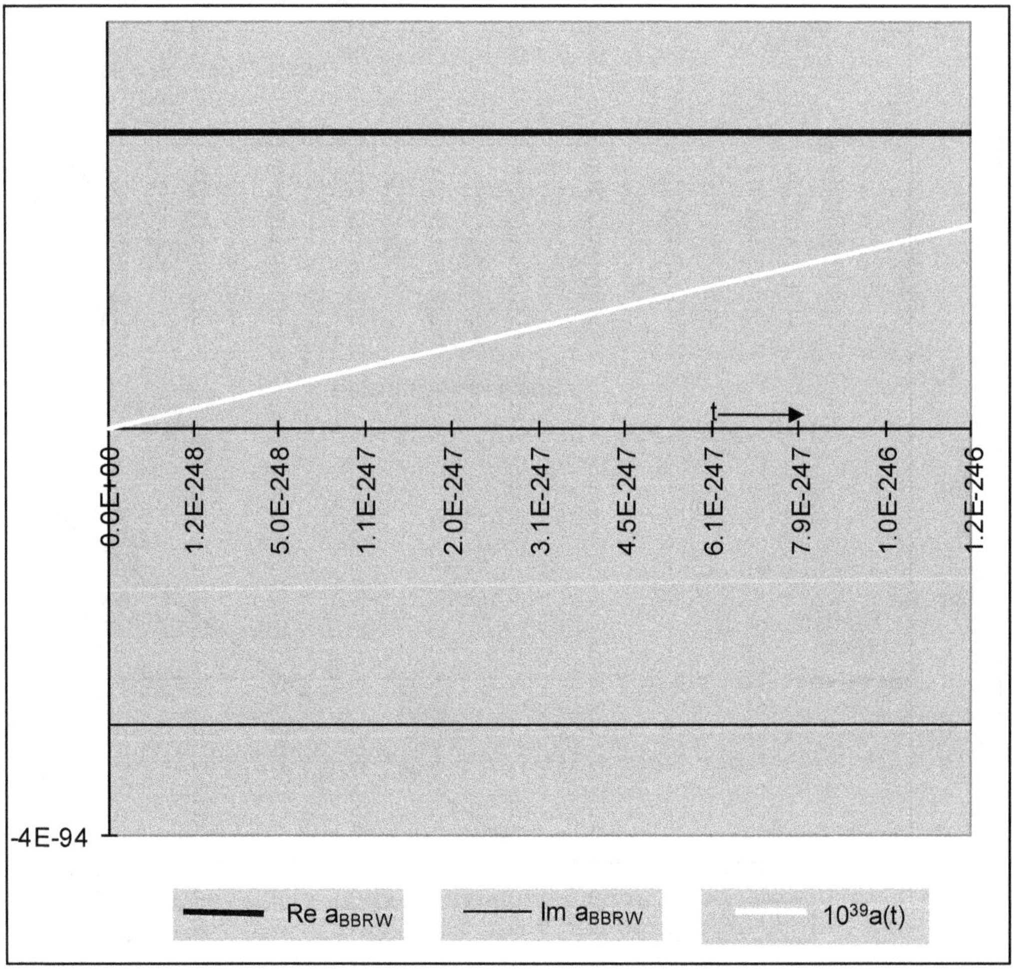

Figure 12.3.5.2. A plot of the real and imaginary parts of a_{BBRW}, and $10^{39} \times a(t)$, versus time from t = 0 to t = 1.2×10^{-246} s. Note they are slowly varying and well behaved in the neighborhood of t = 0 with only a(t) having the value of zero. "E" indicates a power of ten (for example: 2.0E-248 = 2.0×10^{-248} s).

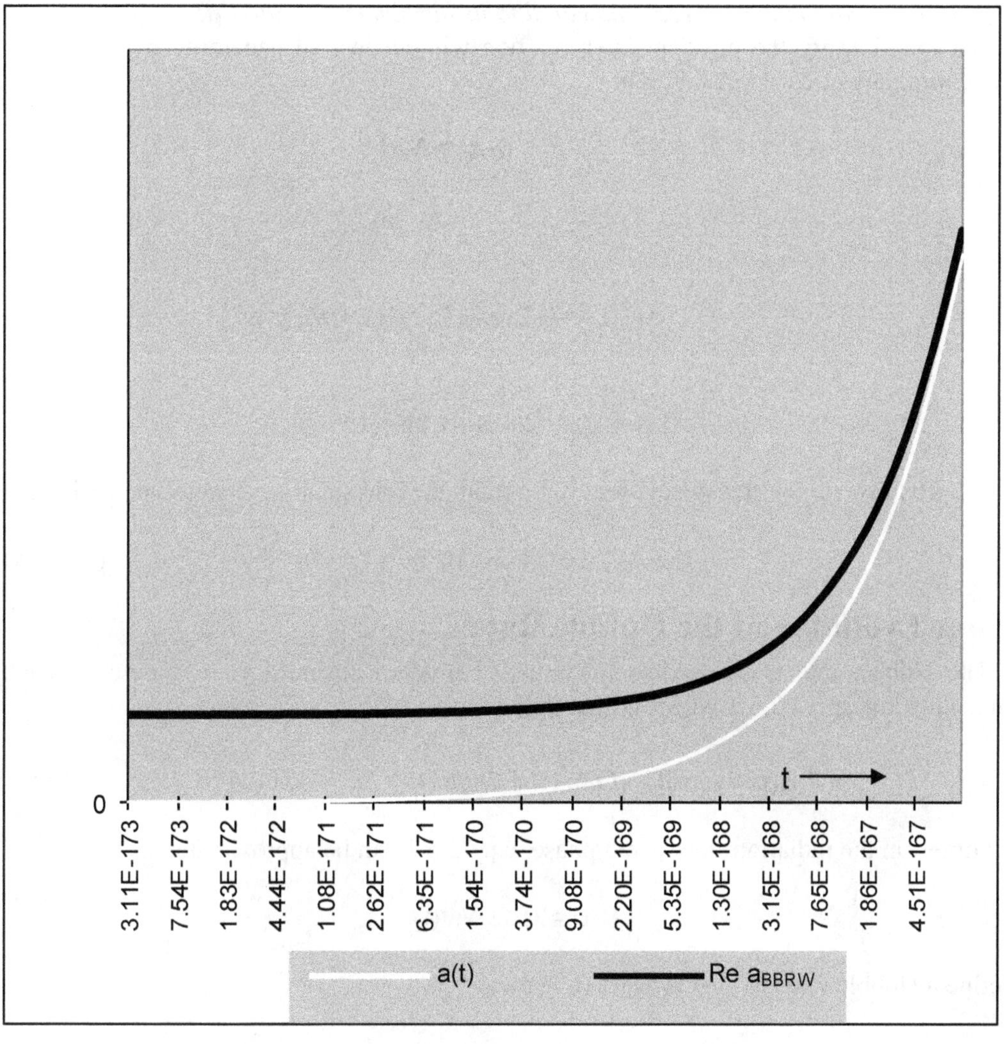

Figure 12.3.5.3. A plot of the real part of a_{BBRW} and a(t) vs. time in seconds around the time 10^{-167} s. Note Re a_{BBRW} quickly changes from slowly growing to growing like a(t).

Having determined the role of the complex scale factor in the metric tensor we can now physically interpret it by applying a Reality group transformation that effectively multiplies $a_{BBRW}(t, \check{r}(r))$ *by a phase making it real and equal to its absolute value* $|a_{BBRW}(t, \check{r}(r))|$.

We can simplify the physical interpretation without loss of generality by considering the radial coordinate of eq. 11-A.4.9 to be

$$r_{GRW} = A_{BB}\check{r} \quad (12.4.3)$$

$$= r_r + ir_i \quad (12.4.4)$$

where

$$A_{BB}(t, y) = 2a_{BBRW}(t, \check{r})/[(1 + M_c^2 y^2)k^{1/2}] \quad (11.5.2.4)$$

Thus

$$r_{GRW} = 2a_{BBRW}(t, \check{r})\check{r}/[(1 + \check{r}^2)k^{1/2}] \quad (12.4.5)$$

Applying a Reality group transformation we obtain the physically measurable, real-valued radius

$$r_{GRWphysical} = 2|a_{BBRW}(t, \check{r})\check{r}/[(1 + \check{r}^2)k^{1/2}]| \quad (12.4.5a)$$

12.5 Time Evolution of the Hubble Rate

The Hubble rate is one of the linchpins of modern cosmology. It is determined by Einstein's equation eq. 12.2.1.1 when written in the form:

$$H(t) = \dot{a}/a = [H_0^2(\Omega_\gamma/a^4(t) + \Omega_m/a^3(t) + \Omega_\Lambda) - k/a^2(t)]^{1/2} \quad (12.5.1)$$

At small times, in the radiation-dominated phase, eq. 12.5.1 can be approximated by

$$H(t) \cong H_0\Omega_\gamma^{1/2}/a^2(t) \quad (12.5.2)$$

If we define a Hubble rate H(t) using $a_{BBRW}(t, \check{r})$ then

$$H_{BBRW}(t, \check{r}) = |\dot{a}_{BBRW}(t, \check{r})/a_{BBRW}(t, \check{r}) \equiv \dot{A}_{BBRW}(t, \check{r})/A_{BBRW}(t, \check{r})|$$

$$= |[H_0^2(\Omega_\gamma/a_{BBRW}^4(t, \check{r}) + \Omega_m/a_{BBRW}^3(t, \check{r}) + \Omega_\Lambda) - k/a_{BBRW}^2(t, \check{r})]^{1/2}| \quad (12.4.3)$$

$H_{BBRW}(t, \check{r})$ is the same as H(t) until we reach the first instants of the universe which we have called the Big Bag Epoch. Then we find

0 ≤ t < t$_c$

$$H_{BBRW}(t, \check{r}) \cong H_0\Omega_\gamma^{1/2}/|a_{BBRW}(t, \check{r})|^2 \qquad (12.4.4)$$

where

$$\text{Re } a_{BBRW}(t, \check{r}) \cong \gamma/2 + a(t)/2 \qquad (12.3.2.24)$$

and

$$\text{Im } a_{BBRW}(t, \check{r}) \cong -\gamma\check{r}/2 - a(t)\check{r}/2 \qquad (12.3.2.25)$$

Substituting in eq. 12.4.4 we find

$$H_{BBRW}(t, \check{r}) \cong 4H_0\Omega_\gamma^{1/2}|[1 - \check{r}^2 + 2i\check{r}]/[(\gamma + a(t))^2(1 + \check{r}^2)^2]| \qquad (12.4.5)$$

$|H_{BBRW}(t, \check{r})|$ is a real physical number in this range. Thus space has a Hubble rate that is both space and time dependent in the Big Bang Epoch.

At t = 0 we find $H_{BBRW}(0, \check{r})$ is finite unlike the radiation-dominated Hubble rate (eq. 12.5.2):

$$H_{BBRW}(0, \check{r}) \cong 4H_0\Omega_\gamma^{1/2}|[1 - \check{r}^2 + 2i\check{r}]/[\gamma^2(1 + \check{r}^2)^2]| \qquad (12.4.6)$$

At the "edge" of the universe the Hubble rate is

$$H_{BBRW}(0, 1) \cong 2H_0\Omega_\gamma^{1/2}/\gamma^2 = 9.5 \times 10^{166}\ \text{s}^{-1} \qquad (12.4.7)$$

and is solely due to the imaginary part of $4H_0\Omega_\gamma^{1/2}[1 - \check{r}^2 + 2i\check{r}]/[(\gamma + a(t))^2(1 + \check{r}^2)^2]$ since the real part is zero.

Thus we consistently avoid the divergences that appear at t = 0 in the Standard Cosmological Model. Its radiation-dominated phase's Hubble rate is $(2t)^{-1}$ which diverges at t = 0.

13. The Big Bang Epoch

No great thing is created suddenly.
Discourses - *Epictetus*

13.1 The t = 0 Big Bang Scenario

The Two Tier cosmological theory that we have developed in preceding chapters differs dramatically from the Standard Cosmological Model in the Big Bang Epoch and yet smoothly melds into the Standard Cosmological Model in the Expanding Universe and Exploding Universe epochs.

The universe has a finite size, temperature and density at the point of the Big Bang which we define to be at the time t = 0. The universe grows slowly for a period of time (until roughly 10^{-167} s) that we call the Big Bang Epoch. In this time period we see a very hot, very dense, macroscopic conglomeration of Y quanta, radiation and elementary particles that coexist with each other with non-singular interactions. These particles are not localized. Each particle can be said to occupy the entire universe since the size of the universe is infinitesimal compared to any particle's Compton radius (if it has one). The universe has an almost classical Robertson-Walker type of metric. In particular, it has a generalized Robertson-Walker metric with quantum corrections due to an effectively classical Y-quanta blackbody radiation field (the inflatons) that is both the source of the metastability of the universe at t = 0 and the source of infinitesimal imaginary spatial dimensions that are comparable in extent with the size of the real spatial dimensions of the universe during the Big Bang Epoch.

As the universe expands due to the Y quanta energy it appears that enormous amounts of gravitational energy are also released since the Two Tier gravitational potential is zero at r = 0 and has a minimum around $r \approx 10^{-33}$ cm. Thus we view the universe in the Big Bang Epoch as in a "slowly" expanding metastable state. (This state is comparable to the metastable false vacuum state in inflation theories – but no scalar bosons are needed – the Y quanta play that role. The combination of gravitation and an effectively classical Y field serve to generate the metastable initial state of the Big Bang.

We can summarize our model's features (many of which are calculated later in this chapter) with:

1. The universe is a macroscopic object in terms of content and as such can be described by classical physics – modified by quantum effects.

2. Quantum fluctuations do not play any significant role in the Big Bang Epoch because of the nature of Two Tier quantum field theory. For example, the quantum fluctuations of the quantized gravitational field were shown to be zero in chapter 7 of Blaha (2004):

$$<0|h_{\mu\nu}(X)h_{\alpha\beta}(X)|0> = \int d^3p\ b'_{\mu\nu\alpha\beta}(p)\ e^{-p^ip^j\Delta_{Tij}(0)}/[(2\pi)^3 2\omega_p] = 0 \qquad (7.3.8.3.3)$$

3. All Two Tier quantum fields also have zero quantum fluctuations. This behavior holds whether they are quantized in flat space or in a curved space. Thus quantum fluctuations (or foam) are not an issue in Two Tier theories.

4. The radiation and matter in the Big Bang Epoch produces, in effect, a classical Y-quanta blackbody spectrum that modifies the nature of space making it complex with the real and imaginary parts of space being comparable. This effect appears in the spatial scale factor of the generalized Robertson-Walker metric described in previous chapters. A Reality group transformation transforms spatial coordinates to real physical values.

5. The usual scale factor a(t) is determined by the conventional Standard Cosmological Model classical Einstein equation since its source is the macroscopic energy density.

6. The radius of the universe (with both real and imaginary parts) at the beginning of the Big Bang Epoch is

$$r_{universe}(0) = a_{BBRW}(0)/k^{1/2} = \gamma(1 - i)/(2k^{1/2}) = 4.3(1 - i) \times 10^{-65} \text{ cm} \qquad (13.1.1)$$

The physical radius is the absolute value of $r_{universe}(0)$. The confinement of particles to a radius of this size means that they cannot be considered to be localized but, rather, they are spread over the entire volume of the universe.

7. The small size of the universe implies Two Tier potentials between particles, and particle propagators, are effectively zero for all particles. Consequently the universe is in a metastable state. Some idea of the relative potential energy of Two Tier QFT particles vs. ordinary QFT particles can be gleaned by comparing a standard Newton-Coulomb type of potential (knowing that it would be modified in strong gravitational fields if they were present)

$$V_{std} = 1/r \qquad (13.1.2)$$

with the Two Tier potential at short distances (below the Planck scale):

$$V_{tt} = 2\sqrt{\pi}\, M_c^2 r \qquad (13.1.3)$$

obtained from eq. 7.3.9.3 of Blaha (2004). We have not displayed the coupling constant in eqns. 13.1.2 and 13.1.3. At $r = r_{universe}(0) = 4.3 \times 10^{-65}$ cm

$$V_{std} = 4.57 \times 10^{59} \text{ ev} \qquad (13.1.4)$$

$$V_{tt} = 2\sqrt{\pi}\, M_c^2 r = 0.00115 \text{ ev} \qquad (13.1.5)$$

Thus the Two Tier potential between particles is negligible at distances up to the radius of the universe at t = 0 since the Heisenberg Uncertainty Principle, when applied using the "diameter" of the universe as the uncertainty in position, implies the uncertainty in a particle's energy, and thus the scale of particle energies, is (coincidentally) of the order of 10^{59} ev. (See the discussion in the following sections.)

8. In view of the above points it is reasonable, and self-consistent, to use the generalized Robertson-Walker metric that we have developed in the preceding chapters.

9. The t = 0 universe is very dense with an energy density of the order of 10^{339} g/cm^3 of particles with negligibly small, non-singular (as particle separation goes to zero) interactions. (See the discussion below.)

10. The Big Bang Epoch is a metastable state. Due to the form of the Two Tier gravitational potential it has a much higher gravitational potential energy at t = 0 than when the radius of the universe is about 10^{-33} cm. **In fact, Gravity is a repulsive force (anti-gravity!) at distances less than 9.08 × 10^{-34} cm.** (Fig. 7.3.9.3 of Blaha (2004) is reproduced on the next page for the reader's convenience.)

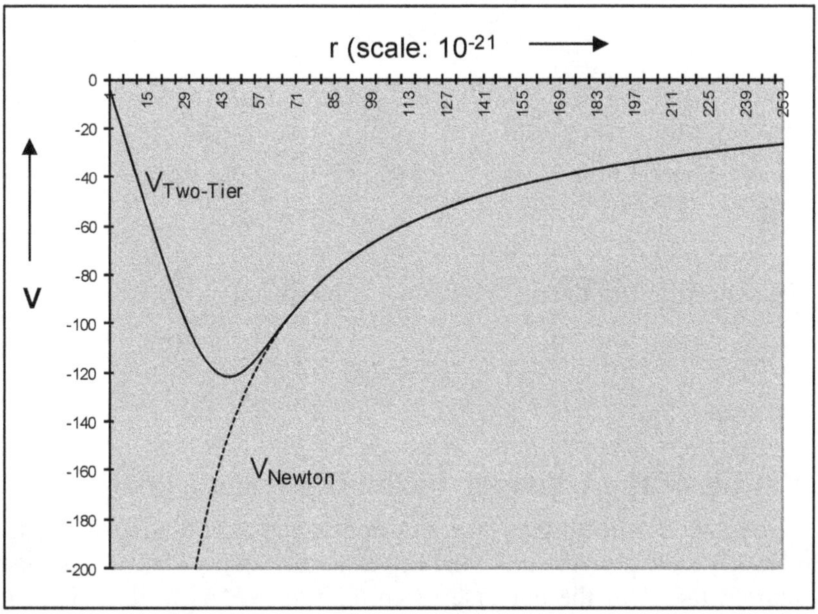

Figure 13.1.1. A plot of the Two Tier gravitational potential (solid line) and the Newtonian gravitational potential (dashed line). Note anti-gravity at short distances. See Fig. 7.3.9.3 of Blaha (2004) for more details.

The Two Tier gravitational potential between two particles has a minimum at (eq. 7.3.9.5 of Blaha (2004))

$$r_{MIN} = \pi^{-½}M_c^{-1} = 9.08 \times 10^{-34} \text{ cm} \qquad (13.1.6)$$

At the minimum the gravitational $V_{\text{Two Tier}}$ has the value:

$$V_{\text{Two TierMIN}} = -.8427\sqrt{\pi}\, GM_c = -1.22 \times 10^{-28} \text{ ev}^{-1} \qquad (13.1.7)$$

Since the energy of the universe is confined within a radius of 10^{-65} cm at t = 0 there is a tremendous release of gravitational potential energy as the universe expands to 10^{-33} cm and beyond. This energy is converted initially into kinetic energy and can be viewed as helping fuel the expansion. A crude approximation to this released energy is

$$\Delta E \sim (\rho V_{\text{universe}})^2\, V_{\text{Two TierMIN}}$$

$$\cong (1.6 \times 10^{212} \text{ g/cm}^3 \times 1.8 \times 10^{-100} \text{ cm}^3)^2 \times 1.22 \times 10^{-28} \text{ ev}^{-1}$$

$$\cong 3 \times 10^{262} \text{ ev} \tag{13.1.8}$$

The energy estimate (eq. 13.1.8) is far beyond the current total energy of the universe which is of the order of 10^{57} ev. Since energy is not conserved (It is considered somewhat undefined by many General Relativists.) as the universe evolves eq. 13.1.8 does not represent a problem in itself.

11. The temperature of the real part of the universe at t = 0 is

$$\kappa T \approx 2 \times 10^{89} \text{ ev} \tag{13.1.9}$$

as shown later.

13.2 The Radius of the Universe in the Big Bang Epoch

In this section we will consider the radius of the universe at the point of the Big Bang (t = 0) and its relatively slow growth during the Big Bang Epoch ($t \le t_c \cong 10^{-167}$ s).

The current radius of the universe (according to WMAP[127] data) is > 7.4×10^{28} cm. Motivated by that finding we have assumed the current radius of the universe in the Robertson-Walker model is twice that value:

Assumption:
$$r_{universe}(t_{now}) = a(t_{now})/k^{1/2} \approx 2 \times 7.4 \times 10^{28} \text{ cm} \tag{13.2.1}$$

and used it to determine k (eq. 12.2.4).

It is reasonable to define the radius of the universe at earlier times correspondingly. In our Two Tier blackbody model for the universe it is

$$r_{universe}(t) = a_{BBRW}(t, \check{r} = 1)/k^{1/2} \tag{13.2.2}$$

with

$$a_{BBRW}(t, \check{r}) = \{-i\gamma\check{r} + a(t) + [\gamma^2 - 2i\gamma\check{r}a(t) + a^2(t)]^{1/2}\}/2 \tag{12.3.2.19}$$

Thus the radius of the universe is complex and given by

$$r_{universeBBRW}(t) = \{-i\gamma + a(t) + [\gamma^2 - 2i\gamma a(t) + a^2(t)]^{1/2}\}/(2k^{1/2}) \tag{13.2.3}$$

The physical radius of the universe is the absolute value of eq. 13.2.3. It is calculated by applying a Reality group transformation.

127 N. J. Cornish, D. N. Spergel, G. D. Starkman, and E. Komatsu, Phys. Rev. Lett. **92**, 201302-1 (2004).

The real and imaginary parts of $r_{universeBBRW}(t)$ are:

$$\text{Re } r_{universeBBRW}(t) = \{a(t) + [R(1 + \cos\psi)/2]^{1/2}\}/(2k^{1/2}) \qquad (13.2.4a)$$

and

$$\text{Im } r_{universeBBRW}(t) = \{-\gamma - [R(1 - \cos\psi)/2]^{1/2}\}/(2k^{1/2}) \qquad (13.2.4b)$$

where

$$R = [(\gamma^2 + a^2(t))^2 + 4\gamma^2 a^2(t)]^{1/2} \qquad (12.3.2.12)$$

and

$$\cos\psi = (\gamma^2 + a^2(t))/R \qquad (12.3.2.13)$$

From the behavior of $a_{BBRW}(t)$ displayed in Figs. 12.3.5.1 – 12.3.5.3 we see that it is well approximated by eqns. 12.3.2.24 – 12.3.2.29 with $\check{r} = 1$. Therefore for $t < t_c$

$t < t_c$

$$\text{Re } r_{universeBBRW}(0 \le t < t_c) \cong (\gamma + a(t))/(2k^{1/2}) \qquad (13.2.5a)$$

$$\text{Re } r_{universeBBRW}(0) \cong \gamma/(2k^{1/2}) = 4.3 \times 10^{-65} \text{ cm} \qquad (13.2.5b)$$

$$\text{Im } r_{universeBBRW}(0 \le t < t_c) \cong -(\gamma + a(t))/(2k^{1/2}) \qquad (13.2.6a)$$

$$\text{Im } r_{universeBBRW}(0) \cong -\gamma/(2k^{1/2}) = -4.3 \times 10^{-65} \text{ cm} \qquad (13.2.6b)$$

from eqns. 12.3.2.24 and 12.3.2.25. Since the Planck length is 1.61×10^{-33} cm we see the real part of the radius of the universe at $t = 0$ (until $t \approx t_c$) is over thirty orders of magnitude smaller than the Planck length.

If we use eqns. 12.3.2.28 and 12.3.2.29 then the radius at present is

$t = t_{now}$

$$r_{universeBBRW}(t_{now}) = a(t_{now})/k^{1/2} - i\gamma/k^{1/2} \qquad (13.2.7)$$

with

$$\text{Re } r_{universeBBRW}(t_{now}) = a(t_{now})/k^{1/2} = 1.5 \times 10^{29} \text{ cm} \qquad (13.2.8)$$

as above (eq. 13.2.1), and

$$\text{Im } r_{universeBBRW}(t_{now}) = -\gamma/k^{1/2} = -8.6 \times 10^{-65} \text{ cm} \qquad (13.2.9)$$

The ratio of the current, and the $t = 0$ real parts of the, radius of the universe is huge:

$$r_{universe}(t_{now})/(\mathrm{Re}\ r_{universeBBRW}(0)) = 3.4 \times 10^{93} \quad (13.2.10)$$

The real part of the universe has expanded dramatically while the imaginary part has remained almost constant in size. The physical radius of the universe for $t < t_c$ and $t = t_{now}$ are the absolute values of the complex radius values above.

The value of the Robertson-Walker radius coordinate at points *within* the universe is specified by

$$r_{RW}(t, \check{r}) = a_{BBRW}(t, \check{r})r \quad (13.2.11)$$

The coordinate r is related to the ř coordinate by

$$r = 2k^{-1/2}\check{r}(1 + \check{r}^2)^{-1} \quad (11\text{-A.4.11})$$

Thus

$$r_{RW}(t, \check{r}) = 2k^{-1/2}\check{r}a_{BBRW}(t, \check{r})(1 + \check{r}^2)^{-1} \quad (13.2.12)$$

The real and imaginary parts of $a_{BBRW}(t, \check{r})$ are specified in eqns. 12.3.2.24 – 12.3.2.29. Again we note the absolute values of eqns. 13.2.11-13.2.12 are the physical values of the Robertson-Walker radius coordinate.

13.2.1 Localization of Particles at t = 0

The radius of the universe in the neighborhood of t = 0 is approximately 6.08×10^{-65} cm. The particles within that incredibly small universe are still the "wave- particles" that we are familiar with within the framework of quantum mechanics. As such, the position and momentum of the particles must satisfy the Heisenberg Uncertainty Condition:

$$\Delta p \Delta x \geq \hbar \quad (13.2.1.1)$$

where $\hbar$ is Planck's constant divided by 2π. In view of the extraordinary small size of the universe – much smaller than the Compton wavelength of any known massive particle – the "spread" (uncertainty) in a particle's position is set by the radius of the universe:

$$\Delta x \approx 2\ |r_{universeBBRW}| \quad (13.2.1.2)$$

Thus the "spread" in the momentum of a particle (the "size" of the region in momentum space where the Fourier transform of the particle's wave function is large) is

$$\Delta p \approx \hbar/\Delta x \approx \hbar/[2|\mathrm{Re}\ r_{universeBBRW}|] = 1.6 \times 10^{59}\ \mathrm{ev} = 1.3 \times 10^{31} M_{Planck} \quad (13.2.1.3)$$

One can only view the particles in the universe in the neighborhood of $t = 0$ as spread across the entire universe. They are entirely unlocalized within the universe from a quantum viewpoint. Since all forces are non-singular in the very small universe at the beginning of time we can view the particles as co-resident in the same spatial region that constitutes the universe. (Simply put, they interpenetrate each other.) Thus the question of particle horizons and the homogeneity of the universe in the Beginning are irrelevant. *The universe today does not scale down to a micro-universe of the same sort as our universe in the neighborhood of t = 0.*

13.3 A Quantum Big Bang and Evolutionary Theory

At this point we have shown that our Quantum Big Bang theory does not have a singularity at the beginning of the universe. It exhibits the known behavior of the universe since 350,000 years after the Big Bang epoch. Thus the long sought inflaton field turns out to be the quantum field, $Y^{\mu}(y)$, appearing in our definition of quantum coordinates. Remarkably our quantum coordinates also eliminate the divergences that have plagued quantum field theory for almost eighty years and enable calculations in The Extended Standard Model and Quantum Gravity to be divergence free without the cumbersome renormalization techniques that were needed for ElectroWeak theory and would not work for Quantum Gravity.

So our quantum coordinates free The Extwnded Standard Model and Quantum Gravity of infinities, both now, and at the beginning of the universe. Naturally this happy removal of infinities through one simple mechanism is a remarkable result that, we believe, reflects the simplicity of Nature when properly understood.Leibniz's Minimax Principle is realized by Two Tier quantum theory that removes the infinities in The Extended Standard Model, Quantum Gravity, and at the beginning of the universe.

13.4 Is the Expansion Rate Increasing?

The expansion rate of the universe has been thought to be increasing for a number of years. Recently, a group[128] led by Professor Subir Sarkar of the Oxford Physics Dept. has used a much larger set of data - a catalogue of 740 Type Ia supernovae (more than ten times the original sample size) to study the expansion rate issue and claims to have found their data is consistent with a constant rate of expansion. The question of an increasing expansion rate is thus still an open question requiring further experiment to resolve the mater.

[128] J. T. Nielsen et al, Marginal evidence for cosmic acceleration from Type Ia supernovae, Nature Scientific Reports (2016); arXiv:1506.01354 (2015)

14. Expansion of Universe by Absorption of Megaverse Matter

While the picture that we have developed of the evolution of the universe in the preceding chapters may be satisfactory in itself to explain the larger features of its evolution, clouds have recently appeared on the horizon due in large part to new experimental results from WMAP and SDSS as well as other collaborations. As a result the issue of an acceleration in the expansion of the universe as well as the issue of "dark energy" have surfaced to muddle the cosmological situation. In addition questions have arisen on the galactic scale about the force of gravity which has led to the possibility of "dark matter" as well as the possibility of modifications of Newton's law of gravitation at distances of the galactic scale and greater.[129]

One possible avenue to approach these issues is to introduce a new recent phase[130] in the evolution of the universe that causes an acceleration in the expansion of the universe. The picture that we have in mind is based on the complex space that we have been using part of in past chapters. In those chapters we limited $\check{r} \leq 1$ because $r < k^{-1/2}$ in Robertson-Walker universes in a real space-time. However, in the case of a complex space-time with a Robertson-Walker or generalized Robertson-Walker metric the radial coordinate is not limited by $r < k^{-1/2}$.

Thus if we exist in a complex space r and $\check{r}$ are not so limited. So if we posit that our universe started as a clump of mass-energy that expanded with the mass-energy in its imaginary parts seeping into its real part as we did in earlier chapters, then we can consider the additional possibility that once the density of the mass-energy in our universe reaches a low enough point mass-energy might also seep in from the regions beyond its limits $r > k^{-1/2}$. Perhaps the larger "universe"[131] of complex space-time can act as a mass-energy source. If the density of mass-energy in the larger universe is greater than the density in our universe it feeds energy into our universe keeping it at the same level as the density of the larger universe. As a result our universe accretes mass-energy that further fuels its expansion and, indeed, may cause its expansion to accelerate. We may call the result an "exploding universe." This scenario does

[129] Except for the last two sections, this chapter is an unpublished paper (2004).. Since the values of cosmological constants have not changed very much in the past 13 years we use data from 2004. New information appearing in paragraphs 1 and 2 has been added to the original paper and some references to the previous model and the Megaverse inserted..

[130] Perhaps beginning a billion years or so after the Big Bang. The previous model might apply to expansion in the initial years after the Big Bang. The model presented in this chapter might apply afterwards when the universe had become sufficiently large to accrete sizeable amounts of mass-energy from the Megaverse.

[131] The Megaverse is now here posited as the source of mass-energy accretion – not the complex space-time of our universe.

raise the question of leakage of mass-energy from our universe to the larger universe at its early stage when its mass-energy density was huge. This leads us to consider the possibility that our universe is in a form of black hole. We explore this idea later.

14.1 Our Island Universe in Complex Space-time

In section 3.8 we briefly considered the possibility that our universe might be one of many universes scattered throughout complex space-time. In section 10.7 we found our blackbody Y quanta model embodied a transfer of mass-energy from imaginary space to real space. Supposing this feature to reflect a general feature of the dynamics of the universe we will now investigate a form of steady state model in which mass-energy entering real space from imaginary space fuels the expansion of the universe in such a way that the mass-energy density of the real space universe remains constant in time (and close to the critical density.)

We start with the commonly accepted isotropic, homogeneous, real space-time of the Robertson-Walker metric. (The model developed in this appendix is independent of the blackbody Y quanta model developed prior to this point in this book.) We also assume

i) a Schwarzschild empty space metric exists in the complex space-time "outside" the universe and that our universe lies within a Black Hole;

ii) a low density of mass-energy exists throughout the complex space-time. The density is assumed to be low enough to enable the approximate validity of the empty space Schwarzschild metric.

iii) the "outside" density "feeds" mass-energy into our universe at each point of space over long periods of time maintaining a constant density in our universe despite its expansion;

iv) the standard Einstein equations describe the dynamical evolution of the universe.

The model that we will develop is similar in some respects to the Bondi-Gold-Hoyle-Narlikar steady state cosmology.[132] (It is also somewhat similar to the deSitter model although the deSitter model assumes negligible pressures and densities (see Peebles(1993))). The Bondi-Gold-Hoyle-Narlikar cosmology is based on an idea called the *perfect cosmological principle*. The perfect cosmolgical principle took Einstein's cosmological principle – the universe was isotropic and homogeneous – and extended it to include the concept that the universe was also unchanging in time – thus the term *Steady-State Cosmology*.

We will assume the mass-energy density of the universe is unchanging in the large but the universe is evolving and may be viewed as a growing black hole with an internal Robertson-

---

[132] H. Bondi and T. Gold, M.N.R.A.S. **108**, 252 (1948); F. Hoyle, M.N.R.A.S. **108**, 372 (1948); F. Hoyle and J. V. Narlikar, Proc. R. Soc. London **A290**, 162 (1966) and references therein.

Walker metric and spatially closed. Thus our model can be viewed as a hybrid of a steady state universe and an evolving universe that starts from a Big Bang. The time period of our model universe begins some time after the Big Bang when the size of the universe is non-zero and the density is finite and fixed during the subsequent evolution. Some interesting features of the model are:

1) The universe always has a finite non-zero size.
2) The Hubble constant changes with time.
3) The universe undergoes an expansion like tha in the de Sitter solution.
4) The age of the universe is approximately 40 billion years (measured from the beginning point of the solution until the present) unlike the Standard Model in which it is of the order of 14 billion years.
5) The universe has a Roberson-Walker metric but is always within a black hole.
6) the Robertson-Walker metric inside the universe is continuous with the Schwarzschild metric outside the universe;
7) The universe is expanding and evolving while maintaining a constant mass-energy density through accretion from the outside complex space.
8) The universe is isotropic and homogeneous.
9) Our model universe may have precursor phases with an inflationary episode as well as a Big Bang period. The universe in the period prior to this model can be assumed to be continuous with the initial state of our model universe.
10) In the 40 billion-year life of our universe it accreted 99.87% of its mass-energy. Thus the prior state of the 0.13% of the mass-energy is swamped and dominated by the overwhelming amount of accreted mass-energy.
11) Since the accreting mass-energy comes from complex space and since a neighborhood of any real space point contains an infinite number of complex space points the accreting mass-energy could homogeneously enter our universe "everywhere" (at a sufficiently low rate.)
12) The exceedingly long life of this model universe may allow for star and galaxy formation so as to account for the abundances of the observed elements. The 70% of the universe callled *dark matter* may be simply burned out stars and galaxies. These are open questions at this time.

14.2 Is the Approximate Equality of the Density of the Universe and the Critical Density a Coincidence?

Occasionally a coincidence in physical phenomena is not a coincidence but a reflection of a principle or deeper physics. A major example of this statement is Einstein's observation that the gravitational mass of an object just happens to equal the inertial mass of an object. This observation was the starting point of the general theory of relativity.

Today we have another amazing coincidence: the observed average density of mass-energy in the universe appears to be just slightly more than the critical density of the universe – the density required for a closed universe. The recent WMAP[133] survey's best estimate (when combined with SDSS data) finds $\Omega_{tot} = 1.012$ with a possible error ranging +0.018 to –0.022 around that value. Ω_{tot} is defined as the

$$\Omega_{tot} = \rho_{universe}/\rho_{cr} \tag{14.2.1}$$

where $\rho_{universe}$ is the mass-energy density of the universe and

$$\rho_{cr} = 3H_0^2/(8\pi G) \cong 9.22 \times 10^{-30}\ \text{g/cm}^3 \tag{14.2.2}$$

is the critical density for closure of the universe with H_0 being the contemporary value of Hubble's constant H. According to the WMAP survey

$$H_0 = 100\text{h km sec}^{-1}\ \text{Mpc}^{-1} = 2.133 \times 10^{-33}\ \text{h ev}/\hbar = \text{h}/9.778\ \text{Gyr}^{-1} \cong 7.56 \times 10^{-29}\ \text{cm}^{-1}$$

with

$$h = 0.70^{+0.04}_{-0.03}$$

and with $\hbar$ being Planck's constant divided by 2π.

Ω_{tot} is approximately unity in the current epoch. Many theories have been developed to explain what can only be described as an amazing coincidence. Generally these theories involve particles that have not been found, or other exotic features, or an incredible fine tuning of parameters in the early universe, or interesting combinations thereof.

We would like to explore a simple model of the recent bulk of the universe's history (excluding perhaps the first few hundred thousand years) that embodies a simple picture of our universe as the interior of an accreting real space region with a Robertson-Walker metric and a constant total density. We assume a Schwarzschild metric describes the complex space-time "outside" the universe. Our universe has real space-time coordinates; the region outside the universe has complex coordinates (i.e. with non-zero imaginary parts).

We assume mass-energy is seeping into our universe from the outer region in just such a way as to keep the average total mass-energy density in our universe constant and approximately equal to the critical density. As our universe accretes mass-energy the universe expands so as to keep the density of the universe a constant value slightly larger than the critical density giving a closed universe. Thus Ω_{tot} = constant ≈ 1 is assumed for much of the lifetime of

[133] M. Tegmark et al, "Cosmological Parameters from SDSS and WMAP", Phys Rev. **D69**, 103501 (2004): Table IV, column 6.

the universe in our model. The period in the history of the universe before our model comes into play may be inflationary or may, perhaps, have an alternate scenario.

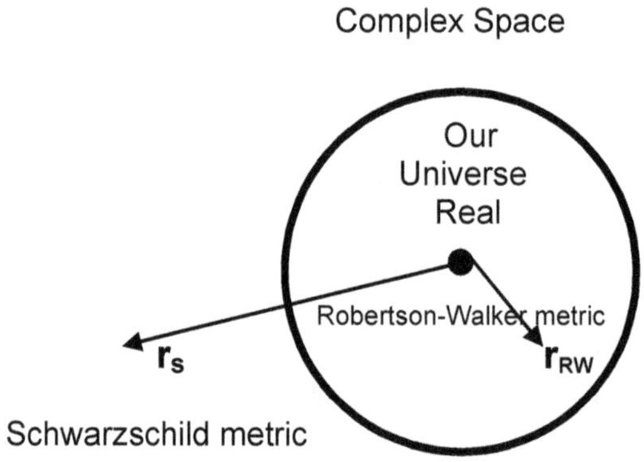

Figure 14.2.1 A visualization of our universe, which can be regarded as the surface of a sphere in a (fictitious) four-dimensional real space. The radius r_{RW} of our Robertson-Walker universe can be related to the radius coordinate r_S of the enveloping complex space-time, which has a Schwarzschild metric.

In this view nothing emerges from the universe. Yet mass-energy enters the universe "everywhere" at an extremely low (undetectable) rate since the neighborhood of every point in our real space-time contains an infinite number of complex space-time points. A very low density of mass-energy is assumed to pervade complex space-time and feed our expanding real space-time universe.

14.3 The Robertson-Walker Metric of Our Universe

We assume our universe is a real subspace of a complex space-time (that is permeated with a very low density of mass-energy); we assume our universe is described by a Robertson-Walker metric (eq. 5.5.1); and we assume the outside complex space-time is approximately described by a complex Schwarzschild metric (eq. 5.2.1.1.1).

We further assume that mass-energy is entering the universe in a homogeneous manner such that the average density is constant in time and slightly larger than the critical density so that space is closed and an outer complex space-time Schwarzschild solution makes sense. Thus

$$\rho \equiv \rho_{\text{universe}} = (1 + \varepsilon)\rho_{\text{cr}} \qquad (14.3.1)$$

is assumed to be independent of time and ε is assumed to a constant with $\varepsilon \ll 1$.

Since ρ is independent of time in this model, $\rho R^3(t)$ is *not* equal to a constant independent of time. Therefore the Einstein equations for Robertson-Walker space-time have the form

$$\dot{R}^2 + k_{RW} = 8\pi G\rho R^2/3 = (1+\varepsilon)H_0^{\,2} R^2 = H'^{\,2}_0 R^2 \qquad (14.3.2)$$

by eq. 14.2.2 where k_{RW} denotes the constant in g_{rr} in the Robertson-Walker metric, where

$$H'_0 = (1+\varepsilon)^{1/2} H_0 \qquad (14.3.3)$$

and where we assume the energy-momentum tensor has the form of a perfect fluid:

$$T_{\mu\nu} = pg_{\mu\nu} + (p+\rho)U_\mu U_\nu \qquad (14.3.4)$$

with p being the pressure and $U_\nu = (1,0,0,0)$ being the velocity 4-vector.

Since the density ρ is a constant in this model one can view eq. 14.3.2 as describing a universe with $\rho = 0$, and a cosmological constant $\Lambda \equiv 8\pi G\rho = 6.76 \times 10^{-66}$ ev^2. We can then rewrite eq. 14.3.2 as

$$\dot{R}^2 + k_{RW} = \Lambda R^2/3 \qquad (14.3.2a)$$

Eq. 14.3.2 can be immediately solved to give[134]

$$R(t) = \tfrac{1}{2}Ak_{RW}e^{-at}/H'^{\,2}_0 + \tfrac{1}{2}A^{-1}e^{at} \qquad (14.3.5a)$$

where A is a constant and where

$$a = H'_0 \qquad (14.3.5b)$$

Notice t =0 is the beginning of the era where the model is assumed to be valid.

There are a number of ways to set a value for A. Some of the possible ways that do *not* work are 1) setting $\dot{R}/R = H'_0$ or H_0 at the present time (this choice requires $A = \infty$); 2) setting $R(0) = 0$ (not possible in a closed universe); or 3) setting $R(0) = 1$ (R(0) has a lower limit as we will see in this model). Perhaps the most reasonable condition to determine A is to require A to be such that the initial value of R(0) is minimized (thus giving the model the maximum range of R values.) Thus we impose the condition:

$$dR(0)/dA = 0 \qquad (14.3.5c)$$

---

[134] We note that *contracting* solutions for R(t) also exist as can be deduced from eq. 5A.3.2.

The solution of this equation gives a value of A that minimizes R(0):

$$A = k_{RW}^{1/2}/H'_0 \qquad (14.3.5d)$$

so that

$$R(t) = k_{RW}^{1/2}\cosh(H'_0t)/H'_0 \qquad (14.3.5e)$$

with the minimum value of R(t) at t = 0:

$$R(0) = k_{RW}^{1/2}/H'_0 \qquad (14.3.5f)$$

and

$$\dot{R}(0) = 0 \qquad (14.3.6)$$

Eq. 14.3.2 implies that R(0) as given in eq. 14.3.5f is the minimum allowed real value of R(t).

The Robertson-Walker constant k_{RW} and ε can be related by evaluating eq. 14.3.2 at t = t_{now} (= the current time) with the result

$$k_{RW} = \varepsilon H_0^2 R_0^2 \qquad (14.3.7)$$

where R_0 is the current value of R(t). Thus the closure of the Roberson-Walker universe requires a positive ε as expected. Eq. 14.3.7 implies

$$R(t) = \varepsilon^{1/2} R_0\cosh(H'_0t) \qquad (14.3.8)$$

The best current SDSS and WMAP survey results give:[135]

$$t_{nowWMAP} = \text{age of universe} = 14.1 \text{ Gyr} \qquad (14.3.9a)$$

$$H_0 = .70/9.778 \text{ Gyr}^{-1} \qquad (14.3.9b)$$

$$\varepsilon = \Omega_{tot} - 1 = .012 \qquad (14.3.9c)$$

It should be noted that $t_{nowWMAP}$ is not independent of the experimentally determined H_0 since it is determined by

WMAP Assumption: $$t_{nowWMAP}H_0 = 1 \qquad (14.3.10)$$

[135] M. Tegmark et al, "Cosmological Parameters from SDSS and WMAP", arXiv:astro-ph/0310723 (October, 2003) (to appear in Phys. Rev. D).

In our model we can calculate a value for t_{now} from eq. 14.3.8 by evaluating it at $t = t_{now}$ and using $R(t_{now}) = R_0$ by definition with the result:

$$1 = \varepsilon^{1/2}\cosh(H'_0 t_{now}) \tag{14.3.11}$$

In our model

$$t_{now}H'_0 = 2.90 \tag{14.3.11a}$$

Thus in our model the universe is much older than the current estimate (eq. 14.3.9a):

$$t_{now} = \text{arccosh}(\varepsilon^{-1/2})/H'_0 = 40.32 \text{ Gyr} \tag{14.3.12}$$

Therefore the recent controversy concerning the apparently small lifetime of the universe, which was determined directly from eq. 14.3.10, is avoided by our model. Observed stars with very long lifetimes fit very nicely into our model universe. We will use t_{now} as the lifetime of the universe in the remainder of this appendix. We note that this value appears to be ruled out by the WMAP analysis, which sets an upper limit of the order of approximately 20 Gyr.

The expansion of our model universe from $t = 0$ until $t = t_{now}$ can be expressed as the following ratio:

$$R(t_{now})/R(0) = \cosh(H'_0 t_{now}) = 9.11 \tag{14.3.13}$$

The pressure p in this model can be obtained directly from the time-time component of the Einstein equations:

$$3\ddot{R} = -4\pi G(\rho + 3p)R \tag{14.3.14}$$

Since

$$\ddot{R} = H_0'^2 R \tag{14.3.15}$$

by eq. 14.3.8 we find

$$p = -\rho = -(1 + \varepsilon)\rho_{cr} \tag{14.3.16}$$

using eq. 14.3.1. Thus the pressure is constant and negative (thus "outward" directed), and equal in magnitude to the critical density. Thus our model is as if it described an ideal fluid with negative pressure where the work performed to increase the volume exactly cancels the effects of the larger volume.

The time component of the energy-momentum conservation law:

$$T^{\alpha\beta}{}_{;\beta} = 0 \quad (14.3.17)$$

for a perfect isotropic fluid implies

$$\dot{p}R^3 = d[(\rho + p)R^3]/dt \quad (14.3.18)$$

The other components of the energy-momentum conservation law are automatically satisfied. Eq. 14.3.16 implies our model is *consistent with the energy-momentum conservation law* as expressed by eq. 14.3.18. *The density and pressure are self-consistently constant.*

The proper spatial volume of our closed Robertson-Walker universe universe is

$$V_3(t) = 2\pi^2 R^3(t) k_{RW}{}^{-\frac{3}{2}} = 2\pi^2\, H'_0{}^{-3} \cosh^3(H'_0 t) \quad (14.3.19)$$

Using the preceding numerical values we find the current value of the volume of our model universe is

$$V_{30} = V_3(t_{now}) = 3.39 \times 10^{88}\ \text{cm}^3 \quad (14.3.20)$$

while the value of the volume at $t = 0$ is

$$V_3(0) = 4.48 \times 10^{85}\ \text{cm}^3 \quad (14.3.21)$$

showing the initial size of the universe in this model of its recent era was also quite large.

The Hubble rate in our model is

$$H(t) \equiv \dot{R}/R = H'_0 \tanh(H'_0 t) \quad (14.3.22)$$

The Hubble rate is time dependent and increases with time. Eq. 14.3.6 implies the model begins with zero Hubble rate (growth):

$$H(0) = 0 \quad (14.3.23)$$

The model gives the current experimental result for the current time t_{now}:

$$H(t_{now}) = H_0 \quad (14.3.24a)$$

$$H(\infty) = H'_0 \quad (14.3.24b)$$

Recently astrophysical experiments[136] have suggested that the rate of expansion is increasing. The rate of expansion in our model is increasing:

$$a_{Hubble}(t) \equiv \dot{H} = H'_0{}^2/\cosh^2(H'_0 t) = (1 + \varepsilon)K_3(t) \quad (14.3.25)$$

using eqs. 14.3.22, 14.3.7 and 14.3.8 and the expression for the Robertson-Walker three-dimensional scalar curvature:

$$K_3(t) = k_{RW}/R^2(t) \quad (14.3.26)$$

Thus the rate of change in the Hubble constant is a direct measure of the three-dimensional spatial curvature in our model. The model predicts

$$K_3(t_{now}) = H_0{}^2/\cosh^2(H'_0 t_{now}) = 6.89 \times 10^{-59}\ \text{cm}^{-2} \quad (14.3.27)$$

$$a_{Hubble}(t_{now}) = 6.96 \times 10^{-59}\ \text{cm}^{-2} \quad (14.3.28)$$

and at t = 0

$$K_3(0) = H_0{}^2 = 5.72 \times 10^{-57}\ \text{cm}^{-2} \quad (14.3.29)$$

Thus our model offers an alternative to the concept that a form of "dark energy" is the source of the acceleration of the expansion of the universe.[137]

The dimensionless acceleration parameter in our model is

$$q_0(t) \equiv -\ddot{R}R\dot{R}^{-2} = -\tanh^{-2}(H'_0 t) \quad (14.3.30)$$

At present $q_0(t_{now}) = -1.012$. Asymptotically $q_0(t) \to -1$ as $t \to \infty$. Thus the acceleration parameter in this model indicates perpetual, although decreasing, acceleration into the future.

14.4 The Schwarzschild Metric in the External Complex Space-time

The Schwarzschild metric in the "outer" (i.e. outside our real space-time) complex space-time is

$$d\tau^2 = U(r)dt^2 - U(r)^{-1}\,dr^2 - r^2(d\theta^2 + \sin^2\theta d\varphi^2) \quad (5.2.1.1.1)$$

---

[136] A.G. Riess et al, Astron. J. **116**, 1009 (1998); S. Perlmutter et al, Astrophys. J. **517**, 565 (1999); B. Ratra and P. J. Peebles, Rev. Mod. Phys. **75**, 559 (2003) and references therein.

[137] For example see S. M. Carroll, arXiv:astro-ph/0310342 (2003) or R. Mainini, A. V. Maccio, S. A. Bonometto and A. Klypin, arXiv:astro-ph/0303303 (2003) and references therein.

where

$$U(r) = 1 - (2MG)r^{-1} \qquad (5.2.1.1.2)$$

while the Robertson-Walker metric within our universe is

$$d\tau^2 = dt'^2 - R^2(t')[dr'^2/(1 - k_{RW}r'^2) + r'^2(d\theta'^2 + \sin^2\theta' d\varphi'^2)] \qquad (5.5.1)$$

(Note that we now use "primed" Robertson-Walker coordinates and "unprimed" Schwarzschild coordinates.) The radial coordinates of these metrics can be related using

$$r = r'R(t') \qquad (14.4.1)$$

14.5 Matching Boundary Conditions Between our Universe and the External Complex Space-time

We will denote the boundary between the regions with these metrics in the respective radial coordinates as r_S and r'_{RW} (depicted in Fig. 14.2.1). They are related by

$$r_S = r'_{RW}R(t') \qquad (14.5.1)$$

Both $r_S = r'_{RW}$ are real valued. We require continuity between the exterior and interior metrics at the boundary between the regions, r_S on the Schwarzschild side (r'_{RW} on the Robertson-Walker side). The continuity condition determines the value of the Schwarzschild metric radius coordinate in terms of the mass contained within the universe. The mass density in our model is larger than the critical density for closure of the universe and thus the mass-energy of the universe is confined within it. Mass-energy coming from the exterior passes through a Schwarzschild horizon into our universe where it remains confined "forever."

Since the neighborhood of every real point within our universe contains an infinite number of complex points outside the universe the mass-energy entering from "outside" can appear anywhere within our universe. The influx of mass-energy is very small per unit time per unit area but builds up over time and space to sizeable amounts. The homogeneity of our universe in the large suggests the flux of incoming mass-energy is homogeneously distributed throughout space. Thus our model has a simple mechanism for a growing universe fed by a low, homogeneous density of mass-energy in the outer complex space.

To implement continuous metric values on the boundary between our closed Robertson-Walker real universe and the Schwarzschild metric in the outer complex space-time we choose to map the Roberson-Walker metric to the form of the Schwarzschild metric using eq. 14.4.1, letting $\theta = \theta'$, $\varphi = \varphi'$. As part of the boundary matching process we use an integrating factor to define t in terms of r' and t' in such a way that the cross-term, the "drdt term", is eliminated.[138]

[138] See Weinberg(1972) p. 345 for a mathematically similar example.

The resulting proper interval of the Robertson-Walker metric *in the neighborhood of the boundary* $r' = r'_{RW}$ has the form

$$d\tau^2 = Wdt'^2 - W^{-1}dr'^2 - r'^2(d\theta^2 + \sin^2\theta d\varphi^2) \tag{14.5.2}$$

where

$$W = (1 - k_{RW}r'_{RW}{}^2/R(t')) \tag{14.5.3}$$

Comparing eqs. 5.2.1.1.1 and 5.2.1.1.2, and eqs. 14.5.2 and 14.5.3 leads to the continuity condition:

$$2MG/r_S = k_{RW}r'_{RW}{}^2/R(t') \tag{14.5.4}$$

where M(t') is the mass-energy of our universe:

$$M(t') = \rho V_3(t') = 3\pi(1 + \varepsilon)\cosh^3(H'_0t')/(4GH_0) \tag{14.5.5}$$

by eq. 14.3.19. Thus

$$M(t_{now}) = 3.23 \times 10^{59} \text{ gm} \tag{14.5.6}$$

14.6 An Expanding Universe within a Black Hole with a Growing Schwarzschild Radius Due to Incoming Mass-Energy

While r' is a comoving variable the boundary value condition eq. 14.5.4 can specify the boundary location where the continuity condition is implemented with a value of $r' = r'_{RW}$ that changes with time – in this case because of the growing mass-energy, and changing curvature, of our universe.

Eqs. 14.5.1, 14.5.4 and 14.5.5 determine the location of $r' = r'_{RW}$ to be

$$r'_{RW} = (2MG/k_{RW})^{\frac{1}{3}} = [3\pi(1 + \varepsilon)/(2\varepsilon R_0^2)]^{\frac{1}{3}} \cosh(H'_0t')/H_0 \tag{14.6.1}$$

and shows the value of the Robertson-Walker radius at which the continuity condition is implemented grows with time as the universe accretes matter. The Robertson-Walker radius is a time independent comoving variable. However the radius point at which a boundary condition is implemented can be time dependent.

The value of the Schwarzschild metric radius coordinate at the boundary is

$$r_S(t') = [3\pi(1 + \varepsilon)\varepsilon^{\frac{1}{2}}R_0/2]^{\frac{1}{3}} \cosh^2(H'_0t')/H_0 \tag{14.6.2}$$

by eq. 14.5.1.

Two questions now present themselves: What is the value of R_0? Is the universe inside a black hole? The answers to these questions are interrelated. If the universe is inside the Schwarzschild radius of its enveloping complex space-time, then

$$r_S(t') \leq 2GM(t') \quad (14.6.3)$$

If we substitute the previously derived expressions for r_S and M then the inequality eq. 14.6.3 becomes

$$R_0 \leq 9\pi^2(1+\varepsilon)^2\cosh(H'_0t')/(4\varepsilon^{1/2}) \quad (14.6.4)$$

Therefore our universe exists inside a black hole in this model from t' = 0 if

$$R_0 \leq 9\pi^2(1+\varepsilon)^2/(4\varepsilon^{1/2}) = 207.4 \equiv R_{BH0} \quad (14.6.5)$$

where R_{BH0} is the minimum value for a universe contained inside a black hole. The Schwarzschild radius r_{SBH}, if we set $R_0 = R_{BH0}$, is

$$r_{SBH}(t') = 2GM(t') = 6.26 \times 10^{28}\cosh^3(H'_0t') \text{ cm} \quad (14.6.6)$$

The orders of magnitude of $r_{SBH}(t')$ are roughly consistent with the volume estimates in eqs. 14.3.20 and 14.3.21.

Thus our universe resides within a black hole. We will examine its characteristics as a black hole in a subsequent section.

The choice of $R_0 = R_{BH0} = 207.4$ enables us to obtain numerical values for quantities of interest in our model

$$k_{RW} = 2.95 \times 10^{-54} \text{ cm}^{-2} \quad (14.6.7)$$

$$R(t') = 22.72 \cosh(H'_0t') \quad (14.6.8)$$

$$K_3(t') = 5.72 \times 10^{-57}\text{sech}^2(H'_0t') \text{ cm}^{-2} \quad (14.6.9)$$

The value of the radius coordinate in the Robertson-Walker metric equivalent to the Schwarzschild radius is

$$r'_{RWBH}(t') = r_{SBH}(t')/R(t') \quad (14.6.10)$$

$$= 2.76 \times 10^{27} \cosh^2(H'_0t') \text{ cm} \quad (14.6.11)$$

On the other hand the boundary at which our universe's Robertson-Walker metric is continuous with the Schwarzschild metric of the outer space is given by eq. 14.6.1. This is "the boundary of our universe" and equals

$$r'_{RWuniverse}(t') = 2.76 \times 10^{27} \cosh(H'_0 t') \text{ cm} \qquad (14.6.12)$$

The WMAP Survey data[139] sets a *lower limit* to the surface of last scattering (effectively the "radius" of the universe) of 24 Gpc or 78.2 billion light years or

$$r'_{universeWMAP}(t_{now}) \geq 2.4 \times 10^{28} \text{ cm} \qquad (14.6.13)$$

In a our closed Robertson-Walker model universe the radius of the universe is remarkably close to the WMAP lower limit. We find the current radius of the universe in our model using eq. 14.6.12 is:

$$r'_{RWuniverse}(t_{now}) = 2.52 \times 10^{28} \text{ cm} \qquad (14.6.14)$$

Comparing eqs. 14.6.11 and 14.6.12 we see a growing difference between the Schwarzschild radius and the radius of the universe:

$$r'_{RWgap}(t') = r'_{RWBH}(t') - r'_{RWuniverse}(t')$$

$$= 2.76 \times 10^{27} \cosh(H'_0 t')(\cosh(H'_0 t') - 1) \text{ cm} \qquad (14.6.15)$$

$$r'_{RWgap}(t_{now}) = 2.04 \times 10^{29} \text{ cm} \qquad (14.6.16)$$

In a sense the universe is an "island" within a black hole. In our model we assume a constant net influx of mass-energy from a relatively boundless external source.

The period that is a precursor to the model presumably contains an inflationary stage as well as other stages. We will not address these points in this appendix.

14.7 Rate of Accretion of Mass-Energy by the Universe

Eq. 14.5.5 implies that a_m = the rate of accretion of mass-energy by the universe per unit time per unit volume as a function of time is

$$a_m(t') = V_3^{-1}\, dM/dt' = 3\rho\, H(t') \qquad (14.7.1)$$

by eq. 14.3.22. Currently the rate is

[139] N. J. Cornish, D. N. Spergel, G. D. Starkman, and E. Komatsu, Phys. Rev. Lett. **92**, 201302-1 (2004).

$$a_{m0} = a_m(t_{now}) = 3\rho H_0 \quad (14.7.2)$$

$$= 2.12 \times 10^{-57} \text{ gm cm}^{-3} \text{ sec}^{-1} \quad (14.7.3)$$

Thus in *one year* the rate of mass-energy increase in *one cubic parsec* is *currently*:

$$1.96 \times 10^6 \text{ gm pc}^{-3} \text{ yr}^{-1} = 1.17 \times 10^{30} m_p \text{ pc}^{-3} \text{ yr}^{-1}$$

where m_p is the mass of the proton. This quantity seems quite enormous but a cubic parsec is a large volume. The rate of mass-energy increase per year in *one cubic kilometer* is *currently*:

$$6.67 \times 10^{-35} \text{ gm km}^{-3} \text{ yr}^{-1} = 4 \times 10^{-11} m_p \text{ km}^{-3} \text{ yr}^{-1}$$

which is a miniscule amount. In a hundred billion years a cubic kilometer acquires a mass-energy equal to the mass-energy of four protons *at the current rate* of accretion. At first glance one would think this to be much too small but the universe is of great size and matter clumps together so the rate is not too small. The form of the mass-energy is not specified. It is some combination of matter and radiation. The rate of accretion at the beginning of this model is zero since H(0) = 0.

The mass-energy of the universe is given by

$$M(t') = 4.26 \times 10^{56} \cosh^3(H'_0 t') \text{ gm} \quad (14.7.4)$$

by eq. 14.5.5. It has increased from

$$M(0) = 4.26 \times 10^{56} \text{ gm} \quad (14.7.5)$$

to

$$M(t_{now}) = 3.23 \times 10^{59} \text{ gm} \quad (14.7.6)$$

or by a factor of 757. Thus the mass-energy of the universe at the beginning of the model's applicability was 0.13% of the current mass-energy. The accretion overwhelmingly dominates the mass-energy content of the universe. It is interesting to note that the Cosmic Microwave Background (CMB) constitutes approximately 0.01% of the current total mass-energy of the universe.[140] Thus it constitutes about 8% of the mass-energy in the universe at the beginning of this model's time period.

---

[140] P. J. E. Peebles and B. Ratra, arXiv:astro-ph/0207347 (2002).

14.8 Implications of the Universe's Enveloping Black Hole

One could discuss the question of Hawking radiation in this model but the net influx of mass-energy dominates the model to the point where Hawking radiation is not of great interest. We will look into the issues of the temperature and entropy of our black hole universe in the next section.

14.9 Entropy of the Universe

Bekenstein, Hawking and others[141] have developed a thermodynamic view of a black hole based on an analogy between the first law of thermodynamics:

$$dU = TdS - PdV \tag{14.9.1}$$

and the relation between the energy difference dM of nearby black hole equilibrium states (non-rotating and of total charge zero)

$$dM = (32\pi G^2 M)^{-1}\, dA = (8\pi GM\kappa)^{-1}\, d(\kappa A/(4G)) \tag{14.9.2}$$

where A is the area of the black hole (in units where c = 1):

$$A = 16\pi G^2 M^2 \tag{14.9.3}$$

and where κ is Boltzmann's constant. Eq. 14.9.2 can be obtained by expressing eq. 14.9.3 as a differential relation after regrouping terms and introducing factors based on dimensional analysis. In this picture the temperature of the black hole is

$$T_{BH} = (8\pi GM\kappa)^{-1} \tag{14.9.4}$$

and the entropy is

$$S = \kappa A/4G = 4\pi\kappa GM^2 \tag{14.9.5}$$

In our model these quantities are given by

$$T_{BH} = H_0[6\pi^2(1 + \varepsilon)k \cosh^3(H'_0 t')]^{-1} \tag{14.9.6}$$

[141] J. D. Bekenstein, Phys. Rev. **D7**, 2333 (1973); S. W. Hawking, Phys. Rev **D13**, 191 (1976); W. G. Unruh, Phys Rev. **D14**, 870 (1976); L. Parker, Phys. Rev. **D12**, 1519 (1975); D. G. Boulware, Phys. Rev. **D12**, 1404 (1975); R. M. Wald, Commun. Math. Phys. **45**, 9 (1975); P. Candelas and D. W. Sciama, Phys. Rev. Lett. **38**, 1372 (1977) and references therein.

$$= 2.89 \times 10^{-31}\cosh^{-3}(H'_0t')\ {}^\circ K \qquad (14.9.6a)$$

$$A = 9\pi^3(1 + \varepsilon)^2\cosh^6(H'_0t')/H_0^2 \qquad (14.9.7)$$

$$A = 4.99\ \times 10^{58}\cosh^6(H'_0t')\ cm^2 \qquad (14.9.7a)$$

$$S = \kappa A/4G = 4\pi\kappa GM^2 \qquad (14.9.8)$$

$$= 6.63 \times 10^{107}\cosh^6(H'_0t')\ {}^\circ K^{-1}\ erg \qquad (14.9.8a)$$

At the present time these values are

$$T_{BH0} = 3.82 \times 10^{-34}\ {}^\circ K \qquad (14.9.9)$$

$$A_0 = 2.86 \times 10^{64}\ cm^2 \qquad (14.9.10)$$

$$S_0 = \kappa A_0/(4G) = 3.80 \times 10^{113}\ {}^\circ K^{-1}\ erg \qquad (14.9.11)$$

Thus the entropy of our universe in a black hole in this model is enormous.

The incoming flux of mass-energy through the area of the black hole enclosing the universe is

$$F = A^{-1}\ dM/dt' \qquad (14.9.12)$$

$$= H_0^2[4\pi^2(1 + \varepsilon)^{½}G]^{-1}\cosh^{-3}(H'_0t')\tanh(H'_0t') \qquad (14.9.13)$$

$$= 5.84 \times 10^{-20}\cosh^{-3}(H'_0t')\tanh(H'_0t')\ gm\ cm^{-2}\ sec^{-1} \qquad (14.9.14)$$

At the present time the flux is

$$F(t_{now}) = 7.67 \times 10^{-23}\ gm\ cm^{-2}\ sec^{-1} \qquad (14.9.15)$$

14.10 Cosmic Microwave Background (CMB)

The temperature of the Cosmic Microwave Background is

$$T_{CMB}(t_{now}) = 2.725\ {}^\circ K \qquad (14.10.1)$$

Since the wavelength of radiation changes proportionately to R(t') as the universe expands and thus the frequency and energy changes inversely proportionately to R(t') as the universe expands we expect the temperature of the CMB to also be inversely proportional to R(t'):

$$T_{CMB}(t') = a_{CMB}/R(t') \qquad (14.10.2)$$

Using the current CMB temperature we find

$$a_{CMB} = R_0\, T_{CMB0} = 565.2\ ^\circ K \qquad (14.10.3)$$

Thus at the beginning of the model universe the CMB temperature was a relatively sedate

$$T_{CMB}(0) = 24.88\ ^\circ K \qquad (14.10.4)$$

In terms of energy $T_{CMB}(0)$ corresponds an energy $\kappa T_{CMB}(0) = 2.14 \times 10^{-3}$ ev. Thus the various phases of nucleosynthesis would have preceded the beginning of this model's applicability.

Since the CMB appears to be about 8% of the mass-energy of the universe at t = 0 it could be viewed as simply a relic of the preceding phase of the universe. It remains to be shown that the relic CMB was not distorted by the accreting mass-energy.

14.11 Dark Matter, Dark Energy and Other Phenomena of this Model

Peebles[142] (1993) provides a timetable for the formation of structures and features in the universe as a function of z – the red shift. In our model the red shift is equal to

$$z = R(t_{now})/R(t') - 1 \qquad (14.11.1)$$

in terms of our variables, or alternately,

$$z = 1/a(t') - 1 \qquad (14.11.2)$$

in terms of a common a common notation for the scale factor. We note $a(t_{now}) = 1$.

In the model universe that we have been examining z ranges from 0 (the present) to z = 8.11 about 40.32 billion years ago. An augmented version of Peebles' timetable appears in Fig. 14.11.1 with some additional columns specific to our model universe.

Some interesting features of our model with respect to the timetable of events are:

1. The initial events of the universe (Gravitational Potential Fluctuations and the creation of Spheroids of Galaxies) appear to take place well before the epoch

[142] p. 611.

described by our model. takes place before the epoch in which our model is applicable.

2. **Dark energy appears to account for 70% of the mass-energy density of the universe. In our model 70% of the mass-energy of the current universe was generated within the past 5.68 Gyr ≈ 6 Gyr. The association of these two seemingly independent experimental numbers is not suggested by any other model with which this author is familiar.** This period is only 20% more than the length of the`earth's existence. Could dark energy simply be recently (within the past 6 Gyr) accreted mass-energy that has yet to differentiate itself into radiating structures. Note our model, if dark energy is interpreted as the most recent 70% accreted) has dark energy beginning to arrive at z ~ 0.5.

3. It appears that the period preceding our model created the "skeleton" of many of the structures in the universe, and the period to which we apply our model put "flesh" on these structures.

Cosmological Event	z	t' (Gyr)
Gravitational Potential Fluctuations	$\gtrsim 10^3$	-
Spheroids of Galaxies	~ 20	-
First Engines for Active Galactic Nuclei	$\gtrsim$ 10	-
The Intergalactic Medium	~ 10	-
Beginning of our model Universe	8.11	0
Hubble Telescope Finds 6 Galaxies	8	
Void of Galaxies – Recently Found	~ 7	
Super-Massive Black Hole (Blazer) Found[143]	5.47	
Dark Matter	$\gtrsim$ 5	$\lesssim$ 13.6
Dark Halos of Galaxies	~ 5	~ 13.6
Angular Momentum of Rotation of Galaxies	~ 5	~ 13.6
Population II Stars Created at t' ~ 0 Expiring	~ 5 - 3	~ 14 - 20
Formation of Many Galaxies (a "wall" of galaxies)	~ 4 - 2	~–25
The First 10% of the Heavy Elements	$\gtrsim$ 3	$\lesssim$ 20.3
Cosmic Magnetic Fields	$\gtrsim$ 3	$\lesssim$ 20.3
Rich Clusters of Galaxies	~ 2	~ 24.7
Thin Disks of Spiral Galaxies	~ 1	~ 30.5

143 R. W. Romani et al, "Q0906+6930: The Highest-Redshift Blazer", arXiv:astro-ph/0406252 (2004).

Superclusters, Walls, and Voids	~ 1	~ 30.5
Start of Accretion interpreted as Dark Energy ?	~ .5	~ 34.64
Present Day Universe	0	40.32

Figure 14.11.1 A timetable for the formation of structures and features in the universe with data obtained in part from Peebles(1993).

4. The beginning point of our model (z = 8.11) is suggestively close to the formation of the first engines of galactic nuclei and the intergalactic medium (z ~ 10). The inflow of mass-energy might have been the source of these developments.

5. Population II stars that were created at the beginning t' ~ 0 start expiring at red shifts z ~ 3 – 5. The remnants of these dead stars (and their partial absorption into galactic black holes) may constitute the "dark matter" that coincidentally first appears at z ~ 5. As Peebles[144] remarks, "If baryonic, when did the matter become dark? The amount of diffuse baryonic matter observed at z ~ 3 in damped L*a* systems, ... is comparable to what is seen now in stars ... and does not seem to be as large as the dynamical estimates of what is present now as dark matter ... this has been taken to be a hint that the matter became "dark" during or possibly a long time beforee the assembly of the massive halos." Thus dark matter may be the remnants of the burnt out stars from the first generation of stars in our model that expired in the first 14 – 16 billion years of the model universe.

6. The model does provide a mechanism for an accelerating expansion rate without the introduction of a new force – an increasing expansion rate without "Cosmic repulsion." Effectively the constant mass-energy density plays the role of a cosmological constant.

14.12 Some Issues with this Model

This model has many features that are similar to theories developed of the inflationary period of the universe. The model shares with inflationary models a de Sitter-like exponential growth – however the scale setting the rate of growth in inflationary theories is vastly smaller than the scale setting the rate of growth in our post-inflationary model.

This model may have some validity as an approximate description of the universe after an inflationary period if there is an influx of matter from complex space. The model does not explain the near equality of the density of the universe to the critical density. It simply assumes it to be true due to incoming mass-energy from complex space.

---

[144] p. 617 Peebles(1993).

On the positive side our model describes a much older universe of about 40 billion years so the existence of very old stars is not even a borderline issue. It also avoids anthropic hypotheses – the universe is as it was from the point of view of mass-energy density. It is consistent with a prior Big Bang and gives an expanding universe as long as the incoming "fuel" continues. It offers an alternative to Dark Energy since the rate of expansion is apparently accelerating. It provides a hypothesis for the nature of Dark Matter as the remnants of the first 20 billion years of the life of the universe. *Most importantly, from the point of view of this monograph, it exemplifies a possible interaction between our real space and complex space.*

On the negative side our expanding universe may not be compatible with the observed features of the Cosmic Microwave Background (CMB) since so much mass-energy enters the universe from outside. The energy spectrum of the incoming mass-energy is not known. There are several possible incoming mass-energy spectra and combinations thereof:

1. A spectrum following a Maxwell-Boltzmann distribution.
2. A "random" spectrum.
3. A blackbody spectrum.

Without experimental guidance it is impossible to choose amongst these possibilities.

There are numerous other issues that would need to be explained within the framework of this model:

1. The increasing percentage of younger galaxies as one probes deeper into the universe and thus looks at the universe at much earlier times.

2. Models of stellar and galactic evolution as well as the abundance of the elements in an old universe with a massive continuing influx of mass-energy.

Thus the status of this model can only be viewed as an "illustrative model" in view of the many open questions, and the lack of experimental support for its primary features such as the influx of mass-energy.

14.13 The Problem of Newly Discovered Massive Black Holes and Galaxies

The age of the universe is assumed to be directly related to the Hubble constant H with the result that the age of the universe is currently, commonly, assumed to be roughly 14 Gyr years. Recent WMAP discoveries of massive black holes and well formed large galaxies raise questions about either the current theories of galaxy formation and/or the age of the universe. Most theorists have been inclined to view the problem as an isuue for theories of galaxy formation. However, the recent discovery that the rate of expansion of the universe is accelerating – thus implying the Hubble constant changes with time – combined with the existence of massive black holes and galaxies shortly after the purported beginning of the universe – together suggest that the age of the universe may be far older than currently thought. Thus the large age of the universe in the model – 40.32 Gyr – presented in this chapter may not be out of the question.

A pair of new studies shows that galaxies in the early universe matured more quickly than theorists expected, suggesting the conventional model of galaxy formation needs some serious tweaking.

14.14 Implied Mass-Energy Density of the Megaverse

If the Bondi-Gold-Hoyle-Narlikar steady state cosmology[145] is correct, or a similar cosmology, based on the 'continuous creation of matter' then we can consider the possibility that the 'new' mass-energy originates in the Megaverse.[146] Megaverse mass-energy enters our universe in some fashion – we assume uniformly since every point of our universe has an infinite number of external Megaverse points in any neighborhood.

Given this hypothesis we can estimate the density of mass-energy in the Megaverse using simple geometric considerations. The flow of the energy per unit volume, $a_{m3}(t)$, in our 3-dimensional universe from the (D – 1)-dimensional spatial part of the Megaverse is related to the average mass-energy density of the Megaverse $a_{mMEGA}(t)$ by

$$a_{mMEGA}(t) = (D - 1)a_{m3}(t)/3 \qquad (14.14.1)$$

If the increase of energy in our universe is

$$a_{m3}(t') = V_3^{-1}\, dM/dt' = 3\rho\, H(t') \qquad (14.7.1)$$

by eq. 14.3.22, then currently the rate in our universe is

$$a_{m30} = a_{m3}(t_{now}) = 3\rho H_0 \qquad (14.7.2)$$

$$= 2.12 \times 10^{-57} \text{ gm cm}^{-3} \text{ sec}^{-1} \qquad (14.7.3)$$

The form of the mass-energy is not specified. It is some combination of matter and radiation. The rate of accretion at the beginning of this model is zero since H(0) = 0.

Consequently, we estimate the current Megaverse mass-energy density is

---

[145] H. Bondi and T. Gold, M.N.R.A.S. **108**, 252 (1948); F. Hoyle, M.N.R.A.S. **108**, 372 (1948); F. Hoyle and J. V. Narlikar, Proc. R. Soc. London **A290**, 162 (1966) and references therein.

[146] It seems likely that the early stages of the evolution of the universe (close to the Big Bang) were dominated by the Big Bang and its immediate aftermath. Later when the universe reached an appreciable size, and thus a substantial 'area' in the Megaverse, the accretion of mass-energy from the Megaverse might begin to dominate.

$$\rho_{mMEGA0}(t) = ca_{mMEGA0}(t) = (D-1)ca_{m30}(t)/3 \qquad (14.14.2)$$
$$= 2.12 \times 10^{-57}[(D-1)/3] \text{ gm cm}^{-D}$$

after multiplying by c = 1. This density is to be compared to our universe's estimated mass-energy density:

$$\rho_{universe} \cong \rho_{cr} = 3H_0^2/(8\pi G) \cong 9.22 \times 10^{-30} \text{ g/cm}^3 \qquad (14.2.2)$$

For values of D of order 100 we see the Megaverse energy-matter density is about 26 orders of magnitude lower than that of our universe.

*Universes are islands of mass-energy in a lower mass-energy density Megaverse. Thus it appears reasonable to view the Megaverse (outside of universes) as **flat** to good approximation.*

14.15 A Possible Transition Between Models of the Universe's Growth

If the steady state model presented above is correct then we can take the estimates of section 14.14 as circumstantial evidence for the existence of the Megaverse.Combined with the experimental and theoretical evidence presented earlier in chapter 2 for a Megaverse we suggest there is more to Reality than our universe.

If the early universe model in chapters 11 – 13 holds and the universe transitions to the model universe of this chapter then the age of the universe calculated here:

$$t_{now} = \text{arccosh}(\varepsilon^{-1/2})/H'_0 = 40.32 \text{ Gyr} \qquad (14.3.12)$$

which is based on an extrapolation to t = 0, must be modified to 14 Gyr. The explosive growth of the early universe phase causes a truncation of the overall growth time from 40.32 Gyr to 14 Gyr. The earlier growth phase can be viewed as stemming from the energy of the initial Big Bang. Later, as the universe gets large with a large area, it can then enter a growth phase dominated by the accretion of matter from the Megaverse. The cross over point between these phases (models) can be determined by requiring the combined ages of both models to yield a total age of 14 Gyr as currently believed. See Fig. 14.15.1.

Thus we can now see support for the Megaverse concept in these models.

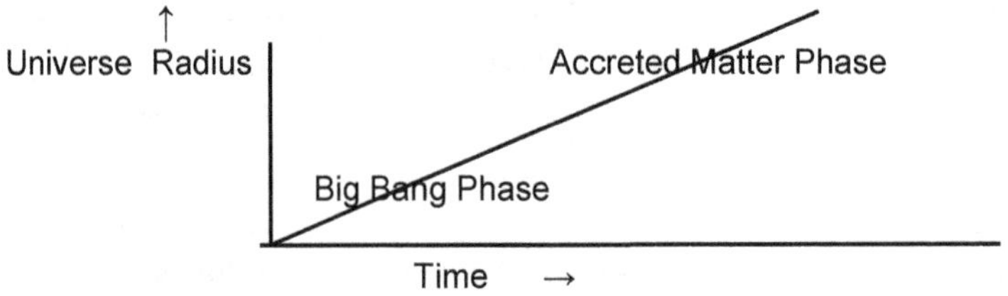

Figure 14.15.1. Rough schematic of universe growth in two phases: Big Bang dominated and Accreted Matter dominated.

15. Why Explore the Megaverse?

This chapter[147] begins the discussion of the entry of Mankind into the Megaverse. Expansion and exploration have been occupations of Man since well before the beginnings of recorded history – indeed since before Man emerged as a species. An examination of history shows that exploration has been of great importance for the growth of Man as a species. Perhaps the most important recent example is the discovery of the Americas which provided not only land for farming and mining but also many new plant varieties (such as corn and potatoes) that have greatly enriched the entire world.

Today Man is beginning the exploration of the solar system.[148] In perhaps a hundred years Man will have the technology and the resources to travel to the stars. While multi-generation starships for slower than light travel are currently the most popular approach, it is our hope – based on our Extended Standard Model theory of elementary particles – that much faster-than-light starships will be built and open the stars for colonization and commerce: The stars – the new America!

After Man has explored and colonized the stars of our galaxy, and then of other galaxies of our universe, the challenge of exploring the Megaverse will present itself. This challenge will not appear soon; but Man will eventually think this universe is too small for his ambitions. Then the Megaverse with its multitude of universes will offer the opportunity to see, and learn from, all the Cosmos. If we must make a guess as to when this challenge will be realized, then it would seem that it will likely be within fifty thousand years, plus or minus, barring scientific developments of an extraordinary nature.

One might ask why we should consider such a remote future which seems to betoken science fiction more than science. The only significant answer to this query is that the knowledge of this distant goal will help guide Man's effort towards its eventual achievement. There is a path within our view. It may not succeed. But it is worthy to consider, and to investigate, over the next millennia. We hope that the challenge of that path will have continuing appeal to subsequent generations of scientists and technologists.

---

[147] This chapter, and the following chapters, are reprinted from Blaha (2014b) and (2014c) with the dimension changed to D.

[148] We have discussed the best approaches to space travel in earlier books: Blaha (2009a) (2009b), (2011b), (2013a). These books have played a significant role in promoting new space initiatives – especially the NASA starship program. The new US Navy heavy electromagnetic guns and rail guns, if extended, can become space guns for the cheap transport of materials from the earth's surface to earth orbit at **enormously lower cost** making space easily accessible to Man. We briefly indicate these possibilities, especially faster than light, quark-gluon ion drive starships later.

As Arnold Toynbee concluded after his monumental study of civilizations, Man can only progress by accepting new challenges and overcoming them. Our study of civilizations[149] conclusively shows the importance of finding 'new land', as Toynbee puts it, with new resources (energy). Simply put: species either grow or stagnate.

15.1 Benefits for Man in Exploring the Megaverse

Most of the semi-intelligent (intelligent?) species on earth that we have encountered have found a niche in Nature which they can dominate. Whales and elephants dominate their niches by their great size. Primates dominate their niches by intelligence and agility – as do dolphins and porpoises. The consequences of a too successful triumph in a niche is stagnation. Whales, for example, have changed very little in the past three hundred million years. Primates have changed because their dominance was not overwhelming and they were confronted with challenges to their survival, which they overcame through change.

Man has achieved dominance over the earth One can foresee a possible future with perhaps new technology but with little change in Man as a species. In fact the great success of medicine has led to a population that appears to be slowly declining physically and mentally on average. Darwinian natural selection has been severely lessened by our humane efforts to cure people, prolong lives, and minister to the handicapped. This may sound callous but growth comes through struggle and the success of the fittest.

The venture into space is an effort to address the challenge of an overcrowded and environmentally degraded earth. One hopes that this effort, if pursued with vigor, will make the planets and moons of the solar system the homes of new civilizations.[150] As in previous history on earth, new civilizations on 'new land' led the way in Man's progress into the future. America is the clear example of this phenomena but one sees it also in the earlier history of Asia and Europe as old civilizations decay and are "replaced" with new civilizations on the frontiers.

The alternative to the growth of civilizations through expansion into 'new land' is a petrified global civilization with a veneer of progress but with a continuation of the current social conditions in a progressively declining planet.[151] The efforts to clean the environment, and improve the earth, in an effectively closed system, such as the earth[152] is, cannot succeed according to the thermodynamic law that entropy always increases in a closed system. An increase in the entropy of the earth is equivalent to a decline in its environment and society.

---

[149] Blaha (2010c) shows energy is required for the development of civilizations.

[150] Otherwise as Professor Hawking has recently said – Mankind's future requires the development of viable colonies on other planets, moons, and lage asteroids. Otherwise the human race faces disaster within 100 years. This author believes the future is not so bleak . Mankind has frequently pulled back from the brink of disaster.

[151] No greater example of the degradation of the earth exists than the large amounts of pollution in the southern Indian Ocean recently seen in the search for a missing Boeing 777 jet – far from the industrialized regions of the world.

[152] The earth is not a completely closed system because it receives energy from the sun. However the energy obtained from the sun will not save earth's environment from degradation as the poisoning of the oceans with mercury from gold mining vividly demonstrates.

These thoughts lead us to consider the distant future, where an analogous situation on a larger scale will exist, when Man, having "conquered" the universe, must accept the challenge of a venture into the other universes of the Megaverse or become a petrified civilization and a declining species.

What benefits can we expect from exploring the universes of the Megaverse? Some of the benefits would be:

1. Encounters with alien civilizations and life that could benefit Man intellectually and technologically just as the Byzantines brought knowledge and civilization to Western Europe leading to a rebirth culminating in the Renaissance.

2. Chemistry and solid state physics are broad areas of study which seem to continually open new possibilities. Thus it is possible that exotic materials and their applications could be found in other universes to the enrichment of Man. Especially interesting is the creation of materials in greater than four dimensions.

3. It is clear from Blaha (2014) that our universe is subject to dynamic forces in the Megaverse. Collisions of universes are possible as well as other calamities. These potential disasters may be much closer in time than the ultimate collapse of the universe, which is often a subject of scientific discussion and news articles. The ability to move part or all of Man into another universe may be a solution to these concerns.[153]

4. It may be possible to develop new senses, or extended senses, by Megaverse exploration from contact with new species or due to new physical environments in other universes. Mankind in the Megaverse might devlop new senses due to the many dimensions of the Megaverse. Mankind adapts to new environments as the Tibetan adaptation to high altitude shows.

5. The successful development of colonies and subsequently human civilizations in other universes would again give man the challenge of a new frontier that could further human progress.

6. Additional unforeseen benefits will be likely. We have seen the technology spinoffs from the space program. The spinoffs from Megaverse exploration should be of a much larger scale.

So we suggest exploration in the Megaverse has multifold advantages for the advancement of Man so that Man, as a species, will not have existed in vain – little better than ants and bees. We should seek to encompass all the Cosmos to extend our understanding to the

[153] Professor Hawking has repeatedly made similar, but less grand, proposals to "save" Man.

entirety of mind and matter. This effort, if successful, will make Man a magnificent new species.

The Anglican Book of Common Prayer states “Death is the Great Victory.” We should like to amend it to “Universal Knowledge is the Great Victory.” That justifies Man as a species.

15.2 Scientific Benefits of Exploring the Megaverse

If we explore the universes of the Megaverse it is quite possible that we will find strange new universes with a different physics and/or a different topology. These potential discoveries would include:

1. Universes with more than four dimensions.

2. Universes with different physical laws.

3. Universes which support extensions of the human senses.

4. Universes with different values for coupling constants and/or masses that might help explain their origin and deepen our knowledge of reality.

5. Universes with a different topology that might sharpen our understanding of General Relativity.

6. The capability to perform comparative studies of universes that will deepen our understanding of space and time. At the moment we are limited to one specimen of the set of universes. Comparative studies of universes would make the study of space and time an “experimental” science.

7. Universes with unique new materials that would substantially improve our technology.

8. The development of a uniship[154] that can enter and traverse the Megaverse would itself generate major advances in our technology (spinoffs) just as the space exploration program has done in the 20th century.

Thus we see that exploring the Megaverse and its universes has broad implications for the far future of Man.

---

[154] A *uniship* is a starship that can enter/exit the universe, and travel in the Megaverse. Perhaps we should have called it a Megaship when we introduced the possibility some years ago.

16. Starships in the Universe and Megaverse

The proof of success in new Science is the benefits that humanity derives and the doorways it opens to a better understanding of Nature. The development of an Extended Standard Model has shed light on the fundamental nature of reality – the nature of space and time, and the basis of elementary particle Physics in Logic – bringing us closer to Plato's vision of reality as essentially a mirror of a World of Ideas.

In this chapter[155] we consider the potential benefits of our work for the future exploration of our universe and the Megaverse in starships capable of enormous speeds and "short" travel times thoughout the Megaverse.

Currently we are exploring the Solar System using relatively slow, unmanned rockets powered by chemical fuels. Soon we will be using nuclear (or fusion) power to explore and colonize selected parts of the Solar System. These approaches to spaceship power are not sufficient to successfully explore and colonize nearby stars. At best they only offer the possibility of multi-generation voyages to even the clsest stars. Multi-generation voyages will be few, and insufficient for colonization and the expansion of humanity to the stars. The fault is basically in the the limitation of travel to speeds below the speed of light. This limitation cannot be overcome with advanced conventional rocketry.

We have proposed the development of a new type of power configuration that will enable starships to far exceed the speed of light using a quark-gluon "ion" drive that *evades* the speed of light limit[156] by accelerating starships with complex-valued thrust. In our derivation of The Extended Standard Model we showed that quarks and gluons have complex speeds in general that are usually masked by color confinement of quarks within hadrons. When heavy atomic nuclei collide in a very high energy accelerator such as CERN in Geneva, Switzerland or the Brookhaven National Laboratory in the United States, a quark-gluon plasma is created that contains free quarks and gluons.[157] These particles, if extracted by magnetic fields before the plasma collapses ("freezes out") and used to provide a complex-valued thrust to a starship,

---

[155] This chapter is reprinted from chapter 25 of Blaha (2015a) with the Megaverse dimension changed to D.

[156] The speed of light is a real number. By using complex-valued speeds a vehicle can "go around" the speed of light to faster than light speeds. Thus Einstein's limit remains true for real velocities. But complex-valued faster than light velocities are possible.

[157] Recently the CERN LHC has studied 7 TeV proton collisions and showed that they produce particles in a manner similar to the particles produced by high energy collisions of atomic nuclei. This enhanced production suggests that we are on our way to producing a quark-gluon ion drive for faster-than-light travel. See J. Adam et al, Nature Physics (2017). DOI: 10.1038/nphys4111. Data from LHC run 1.

would not be limited by the speed of light and could power a starship to large speeds that could reduce travel time to the stars to days rather than hundreds of years.

The quark-gluon drive approach requires very powerful accelerators and magnet technology that have yet to be developed.[158] But our Extended Standard Model theory indicates that it is possible in principle.

We envision a two stage development and exploration effort for the conquest of the universe and the Megaverse realizing that this development effort will take many hundreds of years, and this exploration effort will take tens of thousands of years. Our purpose then is to set a great goal for the world to make a beginning.

The first stage of the effort is to develop accelerator and magnet technology, and other needed technology such as suspended animation, to the point that they can be used to create starship prototypes with faster than light capability and the ability to travel to nearby stars. Subsequently an exploration effort to nearby stars should begin followed by colonization and the exploration of the galaxy. Then jouneys to other far galaxies with an exploration program for interesting parts of the universe.

The second phase of the space initiative would be to travel into the Megaverse and perhaps travel to nearby universes. This phase can be pursued in parallel with phase one after quark-gluon starships become a reality because the doorway to the Megaverse is "just around the corner." In section 16.2 we will a design for a starship that can "slingshot" around a "nearby" neutron star into the Megaverse. Once in the Megaverse the starship, which would have a powerful energy source such as fuel tanks of protons and antiprotons for proton-antiproton annihilation energy and computer support for visualizing and navigation in the D dimensions of the Megaverse, would seek and find a nearby universe for exploration. The mechanism for slingshoting out of a universe would be based on the baryonic force described later. Seeing in the Megaverse could also be based on baryonic planckton r electromagnetic "optics."

After initial exploration, the possibilities of colonization and trade might be pursued if justified. An exploration program for the Megaverse could begin with a planckton observatory searching for other universes just as we now search for galaxies with earth-bound and satellite observatories.

After accomplishing this two phase program Mankind would have achieved its destiny to see all of Creation and to grow to maturity. And yet that will not be the end. For the human mind is an infinite space itself and its exploration will have just begun. In the past ten thousand years, since civilizations began, Mankind has only slightly progressed in its mental abilities and understanding. Much time is needed to grow. See Blaha (2015a).

---

[158] There are also significant other problems such as collisions with space "dust" and debris, and a need for the development of long time hibernation/suspended animation for humans. These problems are solvable.

16.1 Phase One: Starship Engine Development

In this section we will consider a constant, propulsive force in a starship's rest frame that drives the starship from a sublight velocity to a superluminal velocity. The key factor in achieving a superluminal speed is evading the singularity in γ at v/c = 1. We accomplish this goal by having a complex force – a force with a real and imaginary part – that generates a complex acceleration, and thus a complex velocity, that "goes around" the singularity in γ in the complex velocity plane.[159] We assume that an "instantaneous" (Complex) Lorentz transformation relates the earth reference frame and the starship reference frame.

We assume a constant, complex force exists in the rest frame of the starship due to the starship's thrust in the direction of the positive x' (and x) axis. The starship (primed coordinates) and earth (unprimed coordinates) coordinates have parallel axes. The spatial force in the positive x direction is

$$\mathbf{F'} = g\hat{\mathbf{x}} \tag{16.1}$$

where g is assumed to now be a complex constant.

The fourth component of the force (since force is a Lorentz 4-vector) is zero in the rocket's rest frame:

$$F'^0 = 0 \tag{16.2}$$

Applying an inverse Lorentz transformation we find the force in the earth rest frame is

$$\begin{aligned} F^0 &= \gamma(F'^0 + \beta F'^x/c) = \gamma\beta F'^x/c = \gamma v g/c^2 \\ F^x &= \gamma(F'^x + \beta c F'^0) = \gamma F'^x = \gamma g \\ F^y &= F^z = 0 \end{aligned} \tag{16.3}$$

where $\beta = v/c$, c is the speed of light, and $\gamma = (1 - \beta^2)^{-1/2}$ as before. We again use the superscripts x, y, and z to identify the components of the spatial force. The spatial momentum of an object of mass m is

$$\mathbf{p} = \gamma m\mathbf{v} \tag{16.4}$$

and the dynamical equation of motion is

$$d\mathbf{p}/dt = \mathbf{F} \tag{16.5}$$

in the "earth" coordinate system resulting in

[159] This possibility is implicit in the Complex Lorentz group.

$$dp^x/dt = \gamma g \tag{16.6}$$

with[160]

$$dp^y/dt = dp^z/dt = 0 \tag{16.7}$$

The differential equation resulting from eq. 16.5 is

$$d(\gamma v)/dt = \gamma g/m \tag{16.8}$$

Assuming initially, that g is real we must use $\gamma = (1 - \beta^2)^{-\frac{1}{2}}$ for $v < c$ and $\gamma = (\beta^2 - 1)^{-\frac{1}{2}}$ for $v > c$ based on the need for real coordinates for faster than light travel. The solutions for real v are[161]

<u>$v < c$, Re $v_0 < c$</u>

$$v = c\{1 - 2/(1 + ((c + v_0)/(c - v_0))\exp[2g(t - t_0)/(mc)])\} \tag{16.9a}$$

<u>$v > c$, Re $\acute{\upsilon}_0 \geqslant c$</u>

$$v = c\{1 - 2/(1 + ((c + \acute{\upsilon}_0)/(c - \acute{\upsilon}_0))\exp[2\breve{g}(t - t_0)/(mc)])\} \tag{16.9b}$$

where the velocity is v_0 at time t_0, $\acute{\upsilon}_0$ is the velocity[162] at $t = t_0$ and $\breve{g}$ is the acceleration for Re $v \geqslant c$.[163]

Analytically continuing eqs. 16.9 to complex v with a complex force constant g, we obtain the starship equation of motion. We require continuity when the real part of $v = c$ by requiring that when Re v(t) of eq. 16.9a equal c, that $t_0 = t$ and $v(t_0)$ of eq. 16.9a equal $\acute{\upsilon}_0$. These conditions fix t_0 and $\acute{\upsilon}_0$.

<u>Note:</u>
Eqs. 16.9 can easily be integrated to give the distance traveled in the x direction.

<u>$v < c$, Re $v_0 < c$</u>

$$x = x_0 + (mc^2/g)\ln((1 - v_0/c + (1 + v_0/c)\exp[2g(t - t_0)/(mc)])/2) - c(t - t_0 \tag{16.10a}$$

<u>$v \geqslant c$, Re $\acute{\upsilon}_0 \geqslant c$</u>

$$x = x_0 - (mc^2/g)\ln((1 - \acute{\upsilon}_0/c + (1 + \acute{\upsilon}_0/c)\exp[2\breve{g}(t - t_0)/(mc)])/2) - c(t - t_0) \tag{16.10b}$$

or, correspondingly,

---

[160] There is thrust in the y and z direction as well. To avoid getting distracted by the details of an exact calculation we approximate the force in those directions as zero.

[161] The velocity is entirely in the x-direction in this calculation. It can, and does, have complex values in this example.

[162] It is greater than c by assumption in the calculation of eq. 25.9b.

[163] Although eqs. 25.9a and 25.9b have the same form, the acceleration for Re v < c can be changed to a new value ğ after Re v exceeds the speed of light in order to approach the singularity discussed later.

<u>$v < c$, Re $v_0 < c$</u>

$$x = x_0 + (mc^2/g)\ln[(1 - v_0/c)/(1 - v/c)] - c(t - t_0) \qquad (16.11a)$$

<u>$v \geqslant c$, Re $\acute{\upsilon}_0 \geqslant c$</u>

$$x = x_0 + (mc^2/\breve{g})\ln[(1 - \acute{\upsilon}_0/c)/(1 - v/c)] - c(t - t_0) \qquad (16.11b)$$

The complexity of g and thus of the velocity causes x to be complex. The starship is thus generally at a point x in complex space which can be mapped to real-valued space by a Reality grou transformation.

16.1.1 High Speed

To achieve the type of motion we desire the constant force value $\breve{g}$ required after Re $v \geqslant c$ must satisfy a special set of conditions. These conditions emerge from a consideration of the denominator of eq. 16.9b:

$$1 + ((c + \acute{\upsilon}_0)/(c - \acute{\upsilon}_0))\exp[2\breve{g}(t - t_0)/(mc)] \qquad (16.12)$$

where $\acute{\upsilon}_0 \geqslant c$. If this denominator approaches zero then the speed v becomes infinite if g has an appropriate complex value. Let

$$\breve{g} = g_1 + ig_2 \qquad (16.13)$$

If we wish the velocity to get very large (approach infinity) after some acceleration time interval $\Delta t = t_1 - t_0$ we set

$$1 + ((c + \acute{\upsilon}_0)/(c - \acute{\upsilon}_0))\exp[2\breve{g}\Delta t/(mc)] = 0 \qquad (16.14)$$

with the result

$$g_2 = (mc/(2\Delta t))\{n\pi + \text{Im}\ \ln[(c - \acute{\upsilon}_0)/(c + \acute{\upsilon}_0)]\} \qquad (16.15)$$

and

$$g_1 = (mc/(2\Delta t))\ \text{Re}\ \ln[(c - \acute{\upsilon}_0)/(c + \acute{\upsilon}_0)] \qquad (16.16)$$

where n is an odd, positive integer, since $\acute{\upsilon}_0$ is complex in general. Eqs. 16.15 and 16.16 enable the real part of the velocity to become infinite in the time interval Δt. We assume n = 1 in the following discussions. Substituting in eq. 16.9b we obtain

$$v = c\{1 - 2/[1 + ((c + \acute{\upsilon}_0)/(c - \acute{\upsilon}_0))^{1 - (t - t_0)/\Delta t}\, e^{in\pi(t - t_0)/\Delta t}]\} \qquad (16.17)$$

16.1.2 From Light Speed to Enormous Speeds

In the calculation considered above we see an example of accelerating to light speed. In this section we consider the second part of the acceleration: from light speed to enormous speeds taking advantage of the mechanism described in section 16.1.1. We will use an approximation to eq. 16.9b as its denominator approaches zero. Letting $t = t_1 + \tau$ where τ is small, and letting $\Delta t = t_1 - t_0$ then eq. 16.9b becomes

$$\begin{aligned} v &= c\{1 - 2/(1 + ((c + \acute{\upsilon}_0)/(c - \acute{\upsilon}_0))\exp[2\breve{g}(\Delta t + \tau)/(mc)])\} \\ &= c\{1 - 2/(1 - \exp[2\breve{g}\tau/(mc)])\} \\ &\simeq c\{1 - 2/(1 - (1 + 2\breve{g}\tau/(mc))\} \\ &\simeq c\{1 + (mc/\breve{g})(1/\tau)\} \\ &\simeq (\breve{g}^* mc^2/|\breve{g}|^2)(1/\tau) \end{aligned} \tag{16.23a}$$

Continuing the preceding discussion with $\acute{\upsilon}_0 = c + ic$, and m = 10,000 metric tons, and choosing $\Delta t = 30$ days we find

$$\breve{g} = -4.66\times10^{13} + i6.43\times10^{13} \text{ gm-cm/sec}^2 \tag{16.23b}$$

Given the signs of g_1 and g_2 we see that

- For small negative τ both the real and imaginary parts of v approach $+\infty$ as $\tau \to 0$ from below.
- For small positive τ both the real and imaginary parts of v approach $-\infty$ as $\tau \to 0$ from above.

as displayed in Figs. 16.5 and 16.6. A starship can decide to switch off engines and "coast" at high speed towards the destination at some time close to the singularity point.

At $t = 30 - 1\times10^{-13}$ days ($\tau = 8.64\times10^{-9}$ sec) we find, using eq. 16.23a, that the starship velocity in the earth's reference frame is

$$v = 8547c + i11802c \tag{16.23b}$$

which becomes after a Reality group transformation

$$v_r = 14572c \tag{16.23c}$$

At 14572c any of the 100 or so known stars within 21 light years can be reached in a few hours. There is also time needed to decelerate the starship so the actual travel time would be longer. At 30,396c any point in the galaxy could be reached in about 4 years. *Thus Milky Way* *travel times become comparable to 16th century oceanic travel times via ships to various parts of the world!*

16.1.3 The Acceleration Experienced on the Starship

The rapid acceleration, particularly in the neighborhood of $\tau = 0$ raises the question of the inertial forces that would be experienced by passengers on the starship.

The calculation of the maximum acceleration begins with the inverse of the relativistic transformation from earth coordinates to starship coordinates (eq. 16.3):

$$\begin{aligned} F'^0 &= \gamma(F^0 - \beta F^x/c) \\ F'^x &= \gamma(F^x - \beta c F^0) \\ F'^y &= F'^z = 0 \end{aligned} \tag{16.24}$$

which implies the apparent acceleration of the starship calculated in the starship's reference frame is

$$a' = F'^x/m = \breve{g}/m \tag{16.25}$$

whereas in the earth's reference frame, the acceleration a is given by the derivative of eq. 16.9b.

$$\begin{aligned} a &= dv/dt \\ &= 4(\breve{g}/m)((c + \acute{v}_0)/(c - \acute{v}_0))\exp[2\breve{g}(t - t_0)/(mc)]/\{1 + \\ &\quad + ((c + \acute{v}_0)/(c - \acute{v}_0))\exp[2\breve{g}(t - t_0)/(mc)]\}^2 \end{aligned} \tag{16.26}$$

At $t = t_1 + \tau$ we see

$$a \simeq -mc^2/(\breve{g}\tau^2)[1 + 2\breve{g}\tau/mc] \approx -mc^2/(\breve{g}\tau^2) \tag{16.26a}$$

in the earth's reference frame.

The acceleration experienced by the starship occupants, the relevant acceleration for the occupants, is

$$\begin{aligned} a' &= \breve{g}/m = -4.66\times10^3 + i6.43\times10^3 \text{ gm-cm/sec}^2 \\ &= (-4.75 + i6.75)g_E \\ &\equiv 8.25 g_E \end{aligned} \tag{16.27}$$

by eq. 16.25 and 16.23b using a Reality group transformation in the last step. This acceleration (experienced by the starship occupants) is acceptable for short periods of time according to US Air Force studies.

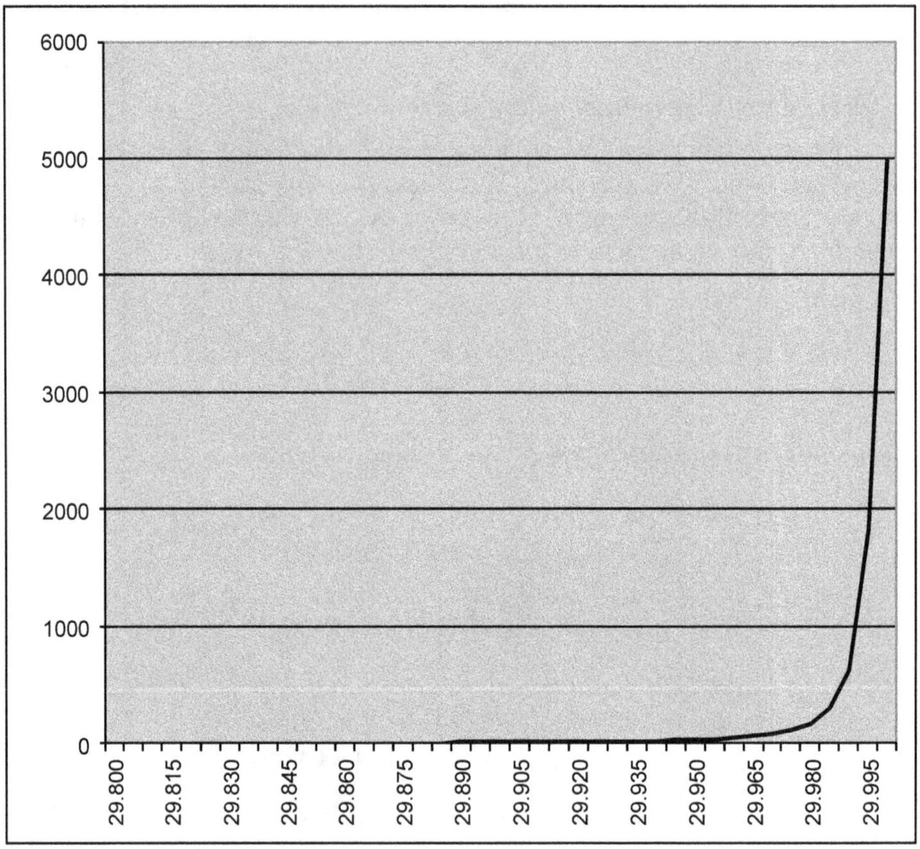

Figure 16.1. A plot of the real part of the velocity of a starship on its 29th and 30th *earth day* of travel up to 5,000c. The dynamics of this case are described in the text where the real speed reaches 14572c and beyond. Time is measured in earth days. Note: as the speed of the starship increases rapidly near the singularity point, time on the starship also passes more quickly so that the starship occupants do not experience very high acceleration. Starship time $t' \approx \beta t$ when $\beta \gg 1$ where t is earth time.

16.1.4 Travel Time on the Starship – Suspended Animation

Another issue is the travel time experienced by the starship occupants. It will appear much, much longer than that measured on earth.[164] For example if v = 5,000c then it will be a factor of approximately 5000 times longer. A 2 month trip from earth's view would take around

[164] Time goes faster at speeds in excess of the speed of light contrary to the sub-light case. See I for the details.

1,000 years from the view of the occupants of the starship. *Therefore a practical method of suspended animation must be found for long distance travel. A 4 month round trip to a star would require the starship occupants to be in suspended animation for approximately 1,700 years – starship time. With suspended animation they could be kept biologically roughly "in sync" with the earth measured travel time of 4 months (plus time spent at the destination) despite the starship elapsed time of about 1,700+ years.*

16.1.5 Constant Superluminal Starship Travel

Having reached an enormous *real* speed such as a speed between 5000c and 30,000c we can turn off the superluminal engines.

The starship then moves at this constant speed in the absence of forces (and neglecting gravity and other minor perturbative forces). At a real speed of 5000c any place in the galaxy is a short travel time away. And nearby galaxies are reachable as well. Figure 16.2 shows the time required to reach various interesting destinations at 30,000c.

Destination	**Distance (ly)**	**Approximate Travel Time (years)**
To the other end of the Milky Way Galaxy	**100,000**	**3**
To the Center of the Milky Way	**30,000**	**1**
Large Magellenic Galaxy	**150,000**	**5**
Small Magellenic Galaxy	**200,000**	**7**
Andromeda Galaxy	**2,000,000**	**70**

Figure 16.2. "Coasting" part of travel time to various destinations at a real velocity of 30,000c.

Since much, much higher "coasting" velocities are also possible, almost the entire visible universe becomes accessible to Mankindgiven sufficient fuel. Mankind then has an incredible future if it has the will to seize it.

16.2 Phase Two: Starship Slingshot into the Megaverse

In this section we will consider a baryonic force mechanism for trips into the Megaverse using a slingshot around a neutron star.

16.2.1 How Can We Evade Euclid's Proof of Three Dimensions and Enter Other Dimensions?

We will now consider a mechanism to escape from a lower dimensional space to a higher dimensional space containing the lower dimensional space as a subspace. We will see that the key factor in this escape mechanism is *a force in the higher dimensional space* that boosts an object in the lower dimensional space into dimensions in the higher dimensional space.

A simple example illustrating this mechanism is a two dimensional flat space (often called *Flatland*) within three dimensional space. We will take the Flatland to be the x-y plane and the z-axis to be the other dimension of the three dimensional space.

Suppose a charged particle is moving with speed v in the positive x direction. Suppose further that a constant magnetic force B points in the positive y direction. Then due to the magnetic vector force law for an electric charge q

$$\mathbf{F} = (q/c)\, \mathbf{v} \times \mathbf{B}$$

the force on the charge will be in the positive z direction – popping the charge out of Flatland into the full three dimensional space. Note that the magnetic force is in the Flatland but the 3-dimensional form of the force law causes the force on the charge to propel it from the Flatland into three dimensional space.[165]

<u>Escaping Our Universe using the Baryonic Gauge Field</u>

The preceding example illustrates the mechanism for particles and uniships to escape from our four dimensional universe into the D-dimensional Megaverse. The force field that will generate the escape is the baryonic gauge field force described in I. It can be used to "pop" baryons and uniships (which are primarily composed of baryons) into the Megaverse.

We shall describe various other methods to provide escape from our universe later.

[165] We assume the escape from the two dimensional subspace is not impeded by a "horizon." We note that this possible impeding horizon does not exist in reality in this ordinary electromagnetism example.

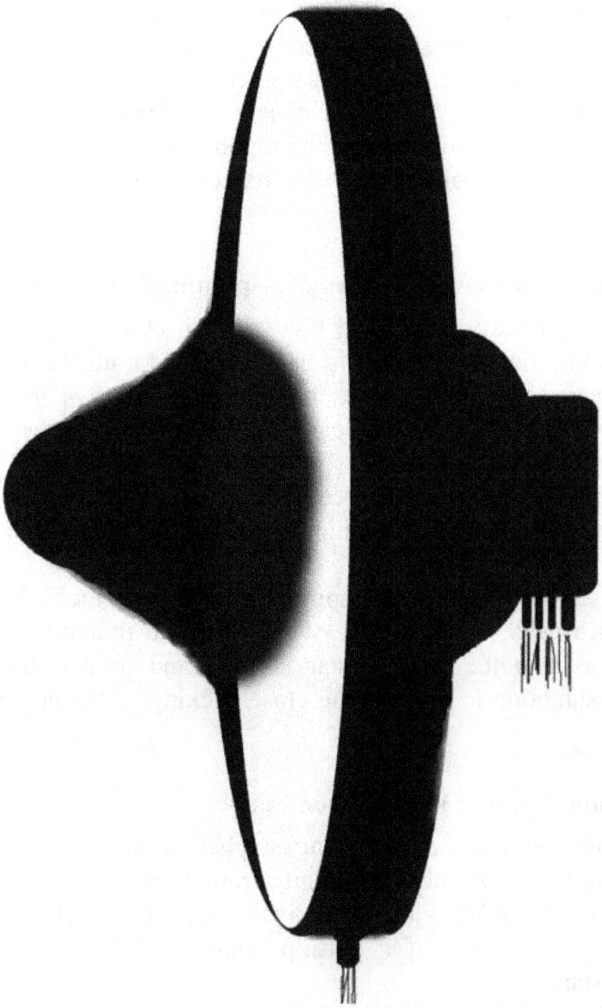

Figure 16.3. A visualization of a starship. The outer disk contains the colliding hadron ring(s) which generate quark-gluon fireballs in a "combustion chamber." The fireball expands through a "rocket nozzle" generating a complex-valued thrust that enables the starship to exceed the speed of light. The four nuclear engines depicted on the underside of the ship provide "intra-solar system" speeds.

An important factor in the nature of neutron stars is their unusually large rotation rates ranging up to over 700 rotations per second. The rotation of a neutron star (assuming a baryonic gauge force exists as we do) generates a *baryonic* magnetic field as well as a baryonic electric field. However we shall see that the neutron star spin-generated baryonic magnetic and electric fields only impart a force to a uniship within our universe. Thus the *Coulomb baryonic force* is the key to slingshoting a uniship into the Megaverse as we pointed out in earlier books.

The important neutron star parameters for the determination of the slingshot trajectory into the Megaverse are its mass, its density as function of radius, and its spin. From these parameters we can determine its baryon number, its baryonic current density, and its baryonic electric and magnetic fields. Then the trajectory of a uniship into the Megaverse can be determined.

16.2.2 The Baryonic Fields of a Rapidly Spinning Neutron Star

Spinning neutron stars, and they all spin at varying rates except in extreme old age, generate baryonic fields which *might* have provided a mechanism to escape from our universe into the Megaverse.[166] In fact, we shall see that *the spin generated force does not* have a component in the direction of the Megaverse spatial coordinates. Thus they do cannot be used to exit into the Megaverse.

The baryonic Coulomb force does have components in Megaverse directions outside our universe and can slingshot uniships (starships designed for travel in the Megaverse) into the Megaverse.

This section calculates the spin-dependent baryonic fields for a neutron star of mass M, internal mass density $\rho(r)$, radius R, volume V, and rotation rate Ω measured in rotations per second. We assume the neutron star is rigid and rotates uniformly. We believe this is a reasonable assumption in view of the close packing of the neutron star throughout most of its body.

16.2.3 Baryonic Current of a Neutron Star

We assume that a neutron star is spherical to good approximation although a rapidly spinning neutron star will deviate slightly from a sphere. We also assume that the neutron star can be treated as point-like since its radius of the order of 10 km is very small compared to the closest point of approach of a uniship which will be, at minimum, tens of thousands of kilometers distant.

The baryonic current of a spinning neutron star, if the coordinate system is oriented so that it spins around the z-axis, yields a current J_φ in the φ direction of spherical coordinates (r, θ, φ). The current is to good approximation

$$J_\varphi = \int dr d\theta d\varphi r^2 \beta_B \Omega \rho(r)/m_n = (\beta_B \Omega/m_n) \int dr d\theta d\varphi r^2 \rho(r) = \beta_B \Omega M/m_n \qquad (16.28)$$

[166] It is possible that other types of stars such as white dwarfs and perhaps even ordinary stars might be used by uniships to escape from a universe into the Megaverse. The enormous mass, and extremely small size, of neutron stars make them more favorable for slingshot escape from the universe.

where β_B is the baryonic charge (analogous to e in electromagnetism), where m_n is the mass of a neutron, $\rho(r)/m_n$ is the neutron (nucleon) number density at radius r, and where the current at a radial distance r is $\beta_B\Omega\rho(r)/m_n$.

As discussed in Blaha (2014) the baryonic "Coulomb" potential in Megaverse coordinates is

$$\phi(y_1, y_2, \ldots, y_{(D-1)}) = (\beta_B{}^2/4\pi)N_1N_2/(y_1{}^2 + y_2{}^2 + \ldots + y_{(D-1)}{}^2)^{1/2} \tag{16.29}$$

where $\alpha_B = (\beta_B{}^2/4\pi)$ is the equivalent of the electromagnetic fine structure constant α. Earlier in I we estimated the order of magnitude of α_B using the differences[167] between various experiments to determine the gravitational constant G. We found the order of magnitude to be

$$\alpha_B = \beta_B{}^2/4\pi \simeq .118\ Gm_H{}^2 \tag{16.30}$$

where m_H is the mass of a hydrogen atom.

We conclude this subsection by determining the baryonic current from the above equations:

$$\begin{aligned} J_\varphi &= (4\pi \cdot 0.118G)^{1/2}\Omega M \\ &= 1.22\ G^{1/2}\Omega M \end{aligned} \tag{16.31}$$

using the approximation $m_H = m_n$.

16.2.4 The D Component Baryonic Vector Potential

Earlier we described some of the features of the baryonic vector potential, which we recapitulate here for the reader's convenience. As in electromagnetism there is an antisymmetric tensor of the second rank that appears in the free part of the baryonic field $F_{Bu\mu\nu}(y)$ lagrangian:

$$\mathscr{L}_{Bu} = -\tfrac{1}{4} F_{Bu}{}^{ij}(y)F_{Buij}(y) \tag{16.32}$$

where

$$F_{Buij}(y) = \partial B_{ui}(y)/\partial y^j - \partial B_{uj}(y)/\partial y^i \tag{16.33}$$

and i, j = 1, 2, … , D. The Dth coordinate corresponds to the time coordinate. While the coordinates are complex in general we will treat the (D – 1) spatial coordinates as real and the Dth coordinate as pure imaginary with the resulting invariant interval

$$ds^2 = dy_1{}^2 + dy_2{}^2 + \ldots + dy_{(D-1)}{}^2 - c^2dy_D{}^2 \tag{16.34}$$

[167] The recent experiment by T. Quinn et al, Phys. Rev. Lett. **111**, 101102 (2013) differs significantly from the 2010 CODATA world average of previous experiments. See P. J. Mohr, B.N. Taylor, and D. B. Newell, Rev. Mod. Phys. **84**, 1527 (2012).. We attribute the difference to the baryonic force between masses.

which is invariant under D-dimensional Lorentz transformations. The coordinates can be transformed into complex-valued coordinates using the Reality group defined earlier.

The tensor F_{Buij} is conveniently separated into a baryon electric part and a baryon magnetic part in a manner similar to the separation of the electromagnetic fields into electric and magnetic fields. However the (D – 1) spatial dimensions change the forms of the baryon fields. Analogously to electromagnetism the baryonic force is given by

$$f_i = F_{Buij}(y)J_B{}^j/c \tag{16.35}$$

where $J_B{}^j$ is the jth baryonic current.

The baryon "electric" field is

$$E_{Bui} = -F_{Bui0}(y)/c \tag{16.36}$$

while the baryon "magnetic" field is

$$B_{Bui} = \varepsilon_{ijk}F_{Bu}{}^{jk}(y) \tag{16.37}$$

where i, j, k = 1, 2, … , (D – 1) and where ε_{ijk} is a totally anti-symmetric tensor with component values ±1. If $i < j < k$ then ε_{ijk} is +1. Even permutations of these three indices yield a value of +1 for the tensor components. Odd permutations of these three indices yield a value of –1. For example, $\varepsilon_{246} = +1$, $\varepsilon_{426} = -1$, $\varepsilon_{642} = -1$, $\varepsilon_{264} = -1$, $\varepsilon_{462} = +1$, $\varepsilon_{624} = +1$.

With these definitions of the $\mathbf{E_{Bu}}$ and $\mathbf{B_{Bu}}$ fields we derive the D-dimensional generalization of the *Lorentz force law* for a baryon of charge q and (D – 1)-velocity v_j:

$$F_i = qE_{Bui} + q\varepsilon_{ijk}v_jB_{Buk}/c \tag{16.38}$$

for i = 1, 2, … , (D – 1). One important difference from the 4-dimensional case is the forms of the $\mathbf{E_{Bu}}$ and $\mathbf{B_{Bu}}$ fields

$$E_{Bui} = -F_{Bui0}(y)/c = [-\partial B_{uD}(y)/\partial y^i - \partial B_{ui}(y)/\partial y^D]/c \tag{16.39}$$

or, expressed as a (D – 1)-vector,

$$\mathbf{E_{Bu}} = [-\nabla_{(D-1)}\phi(y) - \partial\mathbf{B_u}(y)/\partial y^D]/c \tag{16.40}$$

where ϕ is the baryonic Coulomb potential $B_{uD}(y)$, $\nabla_{(D-1)}$ is the (D – 1)-dimensional grad operator, and $\mathbf{B_u}(y)$ is the baryonic (D – 1)-vector potential.

The (D – 1)-dimensional baryon magnetic field has the form of eqn. 16.37. The baryon magnetic field exhibits more complexity than the 3-dimensional magnetic field of electromagnetism:

$$B_{Bu1} = \varepsilon_{1jk}F_{Bu}{}^{jk}(y)/c = [F_{Bu}{}^{23}(y) + F_{Bu}{}^{24}(y) + \ldots + F_{Bu}{}^{2,(D-1)}(y) + F_{Bu}{}^{34}(y) + F_{Bu}{}^{35}(y) + \ldots + F_{Bu}{}^{3,(D-1)}(y) + F_{Bu}{}^{45}(y) + \ldots + F_{Bu}{}^{(D-2),(D-1)}(y)]/c \qquad (16.41)$$

16.2.5 The Baryonic Electric and Magnetic Field Strengths

In this section we will calculate the baryonic electric and magnetic field strengths for a neutron star due to its spin. The spatial field strengths are determined by the dynamical equation:

$$\partial F_{Bu}{}^{ij}(y)/\partial y^i = J^j \qquad (16.42)$$

The current for a neutron star is well approximated by the constant current in the φ direction (in the spherical coordinates)

$$J_\varphi = \beta_B\Omega M/m_n \qquad (16.43)$$

due to the neutron star's small size. Expressing J_φ in rectangular coordinates assuming the rotation is along the z-axis we find

$$J_x = -\sin\varphi\ J_\varphi \qquad (16.44)$$

$$J_y = \cos\varphi\ J_\varphi \qquad (16.45)$$

We will use the relative flatness of space, and the small size of the neutron star neighborhood, to identify the x, y, and z of our universe with the Megaverse coordinates y^1, y^1, and y^3. Inserting eqns. 16.44 and 16.43 in eq. 16.43 yields D – 1 equations:

$$\begin{aligned} \partial F_{Bu}{}^{ix}(y)/\partial y^i &= -\sin\varphi\ J_\varphi \\ \partial F_{Bu}{}^{iy}(y)/\partial y^i &= \cos\varphi\ J_\varphi \\ \partial F_{Bu}{}^{ij}(y)/\partial y^i &= 0 \end{aligned} \qquad (16.46)$$

for j = 4, … , (D – 1). These (D – 1) equations have a solution that gives a magnetic force that is solely within the three spatial dimensions of our universe. *Thus they cannot participate in a slingshot into the Megaverse.*

16.2.6 The Baryonic Coulomb Force Slingshot into the Megaverse

The D-dimensional version of the Lorentz force is

$$F_i = qE_{Bui} + q\varepsilon_{ijk}v_jB_{Buk}/c \qquad (16.47)$$

where

$$\mathbf{E_{Bu}} = [-\nabla_{(D-1)}\phi(y) - \partial\mathbf{B_u}(y)/\partial y^D]/c \qquad (16.48)$$

We have seen that the baryonic force slingshot is wholly derived from the baryonic Coulomb force:

$$\phi(y_1, y_2, \ldots, y_{(D-1)}) = (\beta_B{}^2/4\pi)N_1N_2/(y_1{}^2 + y_2{}^2 + \ldots + y_{(D-1)}{}^2)^{1/2} \qquad (16.49)$$

between two baryon masses with baryon numbers N_1 and N_2. The baryonic Coulomb force is:

$$F_i = \nabla_{(D-1)i}\phi(y) \qquad (16.50)$$

where $\nabla_{(D-1)i}$ is the ith component of the (D – 1)-dimensional grad operator $\nabla_{(D-1)}$.

The part of the Lorentz force that slingshots a uniship into the Megaverse is

$$F_{i\,sling} = \partial\phi(y)/\partial y^i \qquad (16.51)$$
$$= (\beta_B{}^2/4\pi)N_1N_2\, y_i\, /(y_1{}^2 + y_2{}^2 + \ldots + y_{(D-1)}{}^2)^{3/2}$$

for i = 4, 5, … , (D – 1).

A uniship will have a baryonic “Coulomb” force directing it out of our universe into the Megaverse. The baryonic force between the uniship and the neutron star is repulsive since they are both composed of a majority of baryons.

This force will undoubtedly be small compared to the gravitational forces during a slingshot maneuver. Thus we can see that the uniship will slowly ease out of our universe. For a short time it will be partly in and partly out of the universe creating a physical situation not hitherto encountered in physics. It has some advantages since, for example, it allows an “umbrella” of thrust tubes that originally are in 3-dimensional space to widen into a (D – 1)-dimensional umbrella of thrust tubes that would enable the uniship to travel in any direction in the Megaverse – Megaverse maneuverability. We shall consider this possibility later.

Since the force of gravity is confined to within our universe it will have no effect on directions outside our universe. Baryonic forces in directions within our universe will be of little consequence compared to gravitation.

In directions into the Megaverse, gravitation forces being absent, the baryonic force will be the sole force.

Thus we find a clear division: gravitation dominates in directions within the universe; baryonic force dominates in directions out of our universe. Fig. 16.4 depicts the trajectory of a uniship in a slingshot maneuver. Note the attractive hyperbolic motion due to the dominance of gravity in our universe.

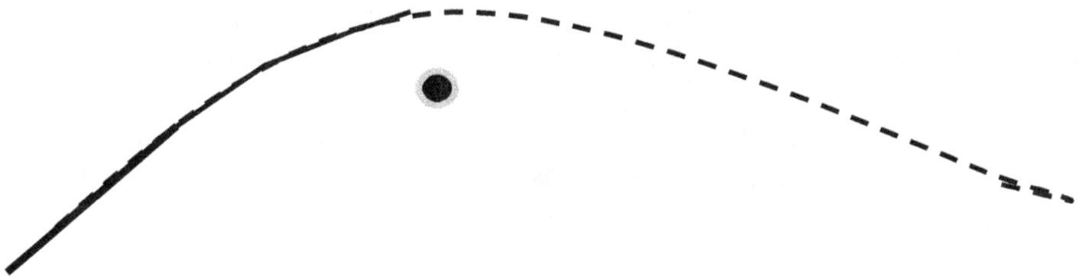

Figure 16.4. The trajectory of a uniship in a slingshot maneuver with a neutron star (the dark circle). The repulsive baryonic force causes the "turn" away from the star into the Megaverse. The solid line corresponds to the time in which the uniship is wholly within the universe and dominated by gravity. The dotted line reflects the transition of the uniship into the Megaverse.

16.2.7 Point-like Uniship Slingshots

The major problems of a close approach of perhaps 100,000 km to a neutron star are the large gravity of the neutron star, its strong tidal gravitation effects (stresses) on the uniship's structure, its strong magnetic field, and its emission of large amount of primarily x-ray/gamma rays. These properties of a neutron star neighborhood would appear to significantly affect the structural integrity of the uniship, and, more importantly, seriously impact on the safety and life of its human crew.

Fortunately there is a saving grace in this physical environment. A uniship approaching the neutron star could resolve these issues by traveling at extremely high speed[168] so that the time spent in the "danger" zone of the neutron star would be very small thus sharply reducing its deleterious effects.

16.2.8 The Uniship Slingshot Trajectory

The slingshot trajectory of a uniship is approximately a hyperbola in our universe due to the dominance of gravity. As it approaches a neutron star, with a distance of closest approach of perhaps 100,000 km, a uniship could be programed to take perhaps 3 seconds or so to circle around the neutron star. Consequently the uniship need only spend a minimal time near the neutron star with little gravitational tidal stress, magnetic field exposure, and radiation exposure issues.

---

[168] The starship would approach at perhaps c/2 and acquire an additional speed due to gravity of c/3. Combining these speeds using the rules of special relativity for the addition of velocities yields an approach speed of 0.7c near the neutron star.

Since the difficulties of a neutron star slingshot are surmountable, we will now turn to the issue of a uniship escape from our universe. We are fortunate that the neutron star's force on the uniship can be conveniently divided into two parts: gravitation which only influences the spatial motion of the uniship in our universe's coordinates, and the baryonic force which only significantly influences the uniship's other D – 4 spatial Megaverse coordinates. The weakness of the baryonic force compared to gravity makes its impact on motion in our universe's spatial coordinates negligible.

The baryonic force generated by the interaction of a uniship's baryons with the baryons within the neutron universe causes the uniship's course to be deflected into the Megaverse.

16.2.9 Uniship Neutron Star Slingshot Dynamics

In this subsection we will describe the dynamics of the neutron star slingshot. We also show that the high gravity of the neutron star causes the universe to have a curvature 'bubble' from flat. This non-zero curvature lowers the 'surface tension' of the universe locally, and thus reduces the confining pressure to a finite value[169] (from infinity), enabling a starship to more easily escape the universe into the Megaverse. The starship goes through the thinner 'fabric' of the universe.

We will assume a flat space-time in our universe with the understanding that space-time may be significantly curved in the immediate vicinity of the neutron star. We assume the uniship will not enter that region. We will define the neutron star to be at the origin of the Megaverse coordinates. Thus $y_i = 0$ for i = 1, … , (D – 1). We will assume the universe spatial coordinates x_i to be equal to the first three Megaverse coordinates:

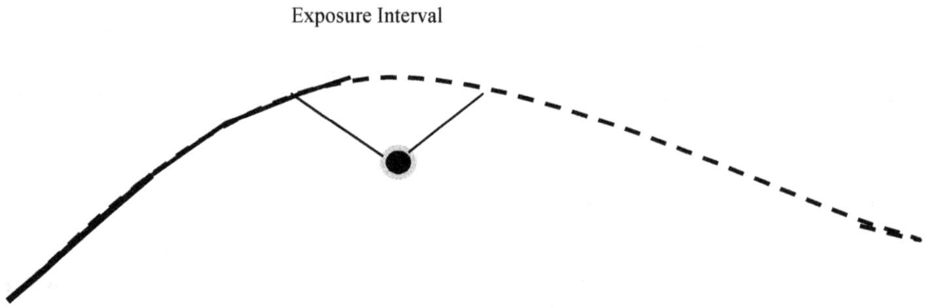

Figure 16.5. Depiction of the uniship exposure interval of closest approach to a neutron star. The time interval spent in this region might be about three second or so. The uniship speed as it approaches might be about 0.7c or so to avoid capture by the neutron star. When the uniship totally exits our universe it disappears from view since electromagnetic radiation (light) from the uniship (or any object) "cannot penetrate" the

[169] See chapter 4 for a detailed discussion of the surface tension and confining pressures of universes.

boundary of our universe.[170] The dashed line indicates the partial exit of the uniship from our universe into the Megaverse with Megaverse velocity and momentum components.

$$y_i = x_i \tag{16.52}$$

for i = 1, 2, 3. We are allowed to do this by the flat space-time assumption for the universe.

The total potential energy of the uniship in the neutron star's reference frame is[171]

$$V_{tot} = -GM_1M_2/r_u + (\beta_B^2/4\pi)\, N_1N_2/r_F \tag{16.53}$$

where M_1 and M_2 are the masses of the neutron star and uniship, N_1 and N_2 are their baryon numbers, and where

$$r_u^2 = x_1^2 + x_2^2 + x_3^2 = y_1^2 + y_2^2 + y_3^2 \tag{16.54}$$
$$r_F^2 = y_1^2 + \ldots + y_{(D-1)}^2 = r_u^2 + y_4^2 + \ldots + y_{(D-1)}^2 \tag{16.55}$$

The force is the gradient of the potential

$$\begin{aligned} F^i_{slingshot} &= \partial V_{tot}/\partial y_i \\ &= GM_1M_2\,(\delta^{i1} + \delta^{i2} + \delta^{i3})y^i/r_u^{3/2} - (\beta_B^2/4\pi)N_1N_2\, y^i/r_F^{3/2} \end{aligned} \tag{16.56}$$

where the Kronecker delta functions restrict the gravity force to the spatial coordinates of the universe.

The dynamic equation of the uniship motion[172] is

$$dp^i/d\tau = F^i_{slingshot} \tag{16.57}$$

where τ is the invariant interval. We now consider the initial phase of the escape to the Megaverse in which

$$r_u >> r_F - r_u \tag{16.58}$$

The universe spatial distance is much greater than the purely Megaverse spatial distance $r_F - r_u$. In this case eq. 16.56 has different forms to good approximation for i = 1, 2, 3 and the remaining Megaverse spatial coordinates:

---

[170] It cannot penetrate in the sense that it is limited by universe coordinates to the confines of the universe. See chapter 8 for details.

[171] Again we note that we are assuming a uniship trajectory in flat space-time so that we may use special relativistic potentials and dynamic equations.

[172] The neutron star is assumed to be stationary due to the largeness of its mass relative to the uniship.

$$dp^i/d\tau \simeq GM_1M_2y^i/r_u^{3/2} \qquad (16.59)$$

for i = 1, 2, 3 and

$$dp^i/d\tau = -(\beta_B^2/4\pi)N_1N_2\, y^i/r_F^{3/2} \qquad (16.60)$$

for i = 4, … , (D – 1). Eq. 16.59 yields a solution of the central force problem which in the present case is an approximately hyperbolic trajectory of the form depicted in Fig. 16.5.

Due to our well-justified assumption that the distance into the Megaverse will be small compared to the distance of the uniship from the neutron star we can approximate

$$r_F^{-3/2} \simeq r_u^{-3/2}[1 - (3/2)(r_F^2 - r_u^2)/r_u^2] \qquad (16.61)$$

Eq. 5.9 then becomes approximately

$$dp^i/d\tau \simeq -(\beta_B^2/4\pi)N_1N_2\, y^i[1 - (3/2)(r_F^2 - r_u^2)/r_u^2]/r_u^{3/2} \qquad (16.62)$$

or

$$dp^i/d\tau \simeq -(\beta_B^2/4\pi)N_1N_2\, y^i[1 - (3/2)\Sigma y_k^2/r_u^2]/r_u^{3/2} \qquad (16.63)$$

where the sum in eq. 5.12 is from k = 4, …, (D – 1). The above equations yield an initially exponential-like trajectory into the Megaverse to leading order.

Thus we have shown that the neutron star slingshot clearly drives the uniship out of the universe into the Megaverse. The uniship passes into the Megaverse. For a small time it is partly in and partly out of the universe.

16.3 Megaverse Surface Tension and a Universe Curvature Bubble

Near a neutron star, or similar small, massive body, the curvature of the universe (which is close to zero – flat) becomes non-zero. Then, as we saw in chapter 4, the surface tension force becomes possibly much smaller (not infinite) and the force keeping mass-energy within the universe decreases enabling a uniship to transit to the Megaverse.

We saw the pressure exerted by the surface tension for a 'spherical' surface area is

$$\Delta p = 2\gamma/R \qquad (4.4)$$

where R equals the radius of curvature of the surface. R = 0 for a flat space. A non-zero value for R causes the infinite force to become a finite force which the uniship can overcome to exit.

Thus the neutron star slingshot has another advantage for transit to the Megaverse. This advantage is enhanced by a faster uniship speed. An extremely large speed causes the uniship to appear 'hot' using the conversion between velocity and temperature. A high speed will create a hotspot in the surface boundary that could substantially lower the surface tension force. Eőtvos showed that the surface tension force has a critical temperature T_c which satisfies:

$$\gamma V^{2/3} = k(T_c - T) \tag{4.5}$$

where k is the Eőtvos constant and V is the volume of the universe (the liquid 'drop'). As the uniship speed increases to an equivalent temperature beyond T_c, the Megaverse surface tension force will drop to zero, and the uniship can freely enter the Megaverse.

Therefore the Megaverse – universe interface can be viewed as a local surface bubbling that thins out at high speed to nothing – no barrier force.

We can estimate the critical velocity v_c as a function of the critical temperature, at which the surface tension force goes to zero, using the velocity maximum of the superluminal Maxwell-Boltzmann distribution:[173]

$$v_c \approx c + ½\, m^2c^5/(2kT_c)^2 \tag{16.64}$$

where m is the mass of the uniship (hundreds of thousands of metric tons presumably). The critical temperature can also be expected to be extremely high. As a result the ratio, the critical velocity, is likely to be large. At sufficiently high superluminal velocity there is no barrier to entry to the Megaverse.

The precise determination of the critical temperature remains to be done. Chapter 14 estimates the critical temperature for our models of chapters 11 - 14.

16.3.1 Umbrella-Shaped Uniship Slingshots

We take it for granted that we can move in one spatial direction or another with ease. However when one enters a higher dimensional space from a space of lower dimension, movement in the additional dimensions, which requires the expenditure of force in those dimensions, becomes an issue. A lower dimension object does not automatically have forces within it in the additional directions and so it cannot move itself, or part of itself in those directions.

In the previous subsections we saw how to use the baryonic force to escape from our universe to the higher dimension Megaverse. In this subsection we will show how to give "wings" to a uniship so that it will have the ability to maneuver in any direction in the Megaverse. To achieve this capability we will have to design the uniship so that it will expand in all Megaverse directions as it enters the Megaverse. This will require an umbrella-like configuration that will use the baryonic force to open the umbrella in all Megaverse 'spatial' directions. The spokes of the umbrella will be long thrust tubes through which uniship thrust can be directed to move the uniship in a desired direction. Fig. 16.6 is a depiction of the simplest form of umbrella uniship.

[173] Blaha (2017c) section 1.4.3.

16.3.2 Scenario for the Opening of a Uniship Umbrella

As an umbrella uniship slingshots around a neutron star the body of the ship including fuel tanks is rigid and moves as a unit. The spokes of the umbrella, the thrust tubes, are moveable and will each move differently because of their differing average distance from the neutron star. As a result they will point in different directions in the Megaverse and can be further moved within the Megaverse relative to each other to deliver thrust in any Megaverse direction. After positioning they can be locked in place and used to maneuver the uniship towards any universe.

16.4 Umbrella-Shaped Uniship Slingshots

We take it for granted that we can move in one spatial direction or another with ease. However when one enters a higher dimensional space from a space of lower dimension, movement in the additional dimensions, which requires the expenditure of force in those dimensions, becomes an issue. A lower dimension object does not automatically have forces within it in the additional directions and so it cannot move itself, or part of itself in those directions.

Earlier we saw how to use the baryonic force to escape from our universe to the higher dimension Megaverse. In this section we will show how to give "wings" to a uniship so that it will have the ability to maneuver in any direction in the Megaverse. To achieve this capability we will have to design the uniship so that it will expand in all Megaverse directions as it enters the Megaverse. This will require an umbrella-like configuration that will use the baryonic force to open the umbrella in all Megaverse directions. The spokes of the umbrella will be long thrust tubes through which uniship thrust can be directed to move the uniship in a desired direction. Fig. 16.6 is a depiction of the simplest form of umbrella uniship.

16.5 Scenario for the Opening of a Uniship Umbrella

As an umbrella uniship slingshots around a neutron star the body of the ship including fuel tanks is rigid and moves as a unit. The spokes of the umbrella, the thrust tubes, are moveable and will each move differently because of their differing average distance from the neutron star. As a result they will point in different directions in the Megaverse and can be further moved within the Megaverse relative to each other to deliver thrust in any Megaverse direction. After positioning they can be locked in place and used to maneuver the uniship towards any universe. (Megaverse navigation is discussed in I.)

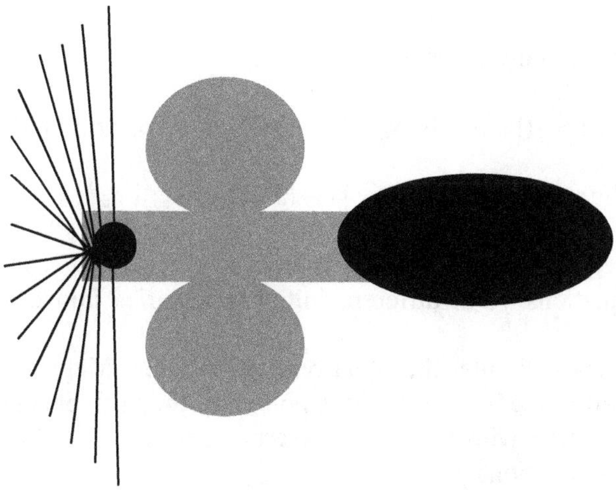

Figure 16.6. Tentative umbrella-like uniship design with the spokes of the umbrella forming a fan. The thrust tubes (D – 1 umbrella spokes – not fully shown) extend kilometers from the thrust power generator(s) core to enable the baryonic force to maneuver them in the (D – 1) different Megaverse directions. The thrust tubes are able to swivel into all (D – 1) spatial directions in response to the baryonic force as the uniship enters the Megaverse. Two fuel spheres are depicted under the assumption that one holds hydrogen and the other holds anti-hydrogen since they are presently the most powerful known possible energy source. The black forward part is for crew, supplies, and cargo. The rear gray part holds the engine apparatus and other related engine components.

16.6 Equations for the Motion of the Thrust Tubes Entering the Megaverse

Earlier we developed the equations for the slingshot mechanism for a compact rigid uniship. In this section we will extend the equations to the case of a rigid uniship with an umbrella with a fan shape (a flat array of spokes as pictured in Fig. 16.6. The array is flat with each spoke in the plane of the ship's trajectory.) A uniship with a true umbrella of spokes (thrust tubes) is another possibility that may be of importance. This is a technical question that we will not address.

We define radial distances from the neutron star center for a "fan-shaped" umbrella with the nth spoke end[174] at radial distance r_{un}. Then we find

[174] The spoke center of mass actually.

$$r_{Fn}^{-3/2} \simeq r_{un}^{-3/2}[1 - (3/2)(r_{Fn}^2 - r_{un}^2)/r_{un}^2] \tag{16.64}$$

for the end of each spoke and we then find

$$dp_n^i/d\tau \simeq -(\beta_B^2/4\pi)N_1N_2\, y^i[1 - (3/2)(r_{Fn}^2 - r_{un}^2)/r_{un}^2]/r_{un}^{3/2} \tag{16.65}$$

or

$$dp_n^i/d\tau \simeq -(\beta_B^2/4\pi)N_1N_2\, y^i[1 - (3/2)\Sigma y_k^2/r_{un}^2]/r_{un}^{3/2} \tag{16.66}$$

for the nth spoke where the sum in eq. 6.3 is from k = 4, …, (D – 1). For each spoke radial distance r_{un} these equations yield different initial trajectories into the Megaverse to leading order.

Thus the spokes will enter the Megaverse in different Megaverse directions since they are not rigidly attached to the uniship body. Upon entry they can change each other's direction giving a final configuration with full Megaverse maneuverability. Uniships can now move in any Megaverse spatial directions.

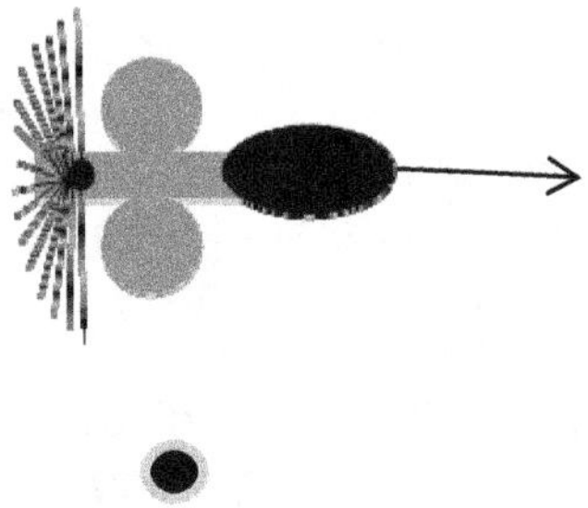

Neutron star

Figure 16.7. Uniship slingshot past neutron star. Note the umbrella spokes which are attached to the uniship but moveable will respond differently to the baryonic force since they are at different radial distances from the neutron star and thus feel differing amounts of force. The baryonic force will twist them in different directions. They then can be re-oriented by the uniship computer to provide mobility in all Megaverse directions.

16.6.1 Equations for the Motion of the Thrust Tubes Entering the Megaverse

Earlier we developed the equations for the slingshot mechanism for a compact rigid uniship. In this subsection we will extend the equations to the case of a rigid uniship with an umbrella with a fan shape (a flat array of spokes as picture in Fig. 16.6. The array is flat with each spoke in the plane of the ship's trajectory.) A uniship with a true umbrella of spokes (thrust tubes) is another possibility that may be of importance. This is a technical question that we will not address.

We define radial distances from the neutron star center for a "fan-shaped" umbrella with the nth spoke end[175] at radial distance r_{un}. Then eq. 16.61 becomes

$$r_{Fn}^{-3/2} \simeq r_{un}^{-3/2}[1 - (3/2)(r_{Fn}^{2} - r_{un}^{2})/r_{un}^{2}] \quad (16.67)$$

for the end of each spoke and then we find

$$dp_n^{i}/d\tau \simeq -(\beta_B^{2}/4\pi)N_1N_2\, y^{i}[1 - (3/2)(r_{Fn}^{2} - r_{un}^{2})/r_{un}^{2}]/r_{un}^{3/2} \quad (16.68)$$

or

$$dp_n^{i}/d\tau \simeq -(\beta_B^{2}/4\pi)N_1N_2\, y^{i}[1 - (3/2)\Sigma y_k^{2}/r_{un}^{2}]/r_{un}^{3/2} \quad (16.69)$$

for the nth spoke where the sum in eq. 16.66 is from k = 4, …, (D – 1). For each spoke radial distance r_{un} these equations yield different initial trajectories into the Megaverse to leading order.

Thus the spokes will enter the Megaverse in different Megaverse directions since they are not rigidly attached to the uniship body. Upon entry they can change each other's direction giving a final configuration with full Megaverse maneuverability. Uniships can now move in any Megaverse spatial directions.

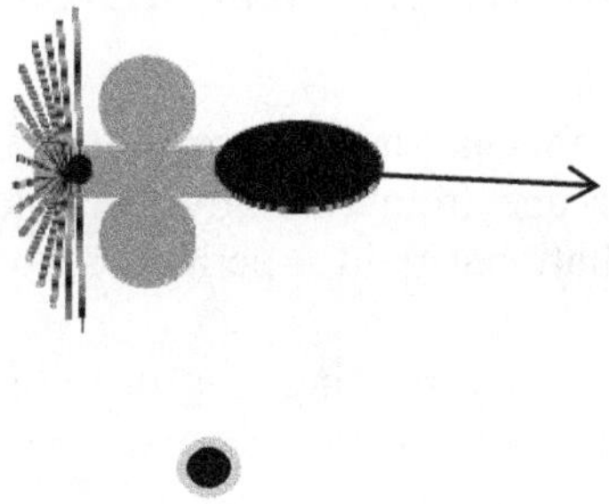

Figure 16.8. Uniship slingshot past a neutron star. Note the umbrella spokes which are attached to the uniship but moveable will respond differently to the baryonic force since they are at different radial distances from the neutron star and thus feel differing amounts of force. The baryonic force will twist them in different directions. They then can be re-oriented by the uniship computer to provide mobility in all Megaverse directions.

175 The spoke center of mass actually.

17. Alternate Approaches to Escaping our Universe

There are two other possible approaches to escaping from our universe into the Megaverse.[176] One approach sends a uniship through a massive ring or cylinder and uses the baryonic force to enter the Megaverse. The other approach uses a rotating uniship, which of course is primarily made of baryons, to rapidly spin relative to a baryon concentration in its neighborhood thus generating a baryonic force that enables the uniship to enter Megaverse.

Both approaches have some merit. But they also have the drawback of requiring large nearby baryon masses of the order of the sun's mass as well as other drawbacks that make them less desirable than the neutron star slingshot approach described in the previous chapter. Again we defer detailed discussion in I.

17.1 Rotating Baryon Ring/Cylinder Mechanism for Escape from Universe

The weakness of the baryonic force requires a large mass, or a gateway ring or a cylinder mechanism to escape from the universe. Alternate mechanisms probably should be located near a large gravitational field – perhaps revolving around a black hole. (See I for an explanation.)

We will now consider a rotating ring or cylinder mechanism consisting of a rotating ring or, more likely, a rotating cylinder of great size and length. Not knowing the strength[177] (coupling constant) of the baryonic force we can only be certain that a baryonic ring or cylinder must be of great size, and large baryonic mass. (Most solid substances are overwhelmingly composed of baryons by mass.) Making a ring, or cylinder, rotate at high rotational velocity will require an extremely large amount of energy. But once set in motion the ring or cylinder will continue to rotate indefinitely. The rotating ring/cylinder must be electrically neutral. The rotating baryon ring/cylinder does have a baryon charge which generates a baryonic field just as a rotating charged ring generates an electromagnetic field.

The baryon numbers required for a baryon ring or cylinder would have to be extremely large (thus large masses) to generate the required sizeable baryonic "magnetic" field. Thus these mechanisms for uniship entry into the Megaverse are at least many thousands of years into the future.

---

[176] This chapter appeared in Blaha (2014c).

[177] We estimate the coupling constant earlier and in I. It is extremely small relative to the coupling constants of other forces.

17.2 Rotating Ring/Cylinder/Mass Configuration/Baryon Cloud Baryonic Force for Escape to Megaverse

A rotating ring, cylinder or mass configuration has a baryon current that generates baryon fields and forces (in a manner analogous to a rotating configuration of electric charges generates electromagnetic fields and forces). This baryon current generates a baryon field that will cause a uniship to spiral out of our universe.

The current in the rotating source creates forces between baryon current loops similar to those of the Biot-Savart law in electromagnetism.

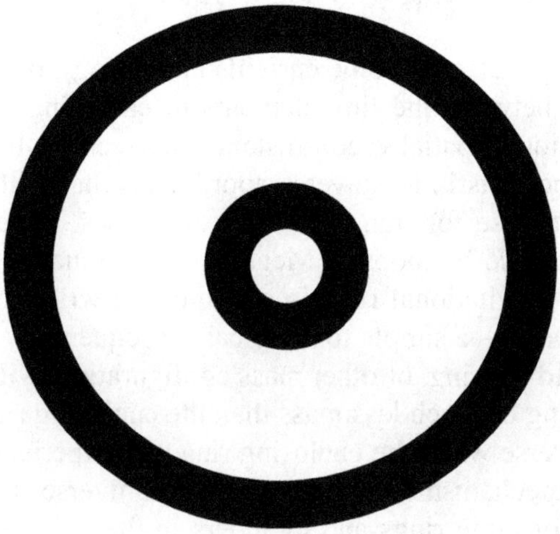

Figure 17.1. A cross section of an outer baryon current loop (in a ring, cylinder, mass configuration, or baryonic cloud) and an inner baryon current loop generated by a large spinning uniship. A "ring within a ring" configuration.

The baryons in the outer "loop" of baryons (ring, cylinder, mass configuration, or baryonic cloud) are effectively fixed in solid masses or slightly moving in a baryon cloud. Thus baryon currents are unlike electromagnetic currents which are generated by moving electrons or ions.

The baryon current generated by the uniship is from the combined rotation of the uniship baryons. It is approximately equivalent to a ring if the uniship is approximately cylindrical.

We have assumed the inner and outer rings are within a region with a large gravitational field such as a region near a black hole or massive star. The presence of a large gravitational field causes the coordinates of the universe x^μ to differ significantly from Megaverse coordinates y^i. Temporarily ignoring quantum aspects, universe coordinates are related to Megaverse coordinates by

$$y_i = f_i(x) \tag{17.1}$$

Because of the disparity in coordinates the baryonic Biot-Savart force implies a baryonic Ampère force between the currents of the form

$$\mathbf{F} = (I_{b1}I_{b2}/c^2) \oint \oint \mathbf{dy_1 \cdot dy_2}\ \mathbf{y}/|\mathbf{y}|^3 \tag{17.2}$$

where I_{b1} and I_{b2} are the Megaverse baryonic currents in the rings, $\mathbf{dy_1 \cdot dy_2}$ is the spatial (D – 1)-dimensional inner product between the line elements of each ring *in Megaverse coordinates*, and **y** is the (D – 1)-dimensional spatial vector distance between the line elements.

F clearly has components in Megaverse coordinates that will propel the uniship out of our universe into the Megaverse for transit to other universes. Note the current loops in our universe's coordinates must also be loops in Megaverse coordinates. These loops which might appear circular in a strong gravitational field in our universe will appear to be distorted closed loops in Megaverse coordinates – a simple topological consequence.

Both the uniship and the ring, or other mass configuration, will experience **F** in opposite directions. The uniship, being of much less mass than the enclosing ring, will experience a large acceleration into the Megaverse while the enclosing ring will experience negligible acceleration.

Thus these other mechanisms for escaping our universe are feasible in principle.[178] However the construction of large rings and cylinders in the neighborhood of a black hole or other large mass is a major technological issue that clearly will not happen (if it does) for many millennia. It requires a new massive scale of technology far beyond our present technology.

The neutron star slingshot mechanism appears to be the most promising approach – especially when one realizes that other universes will not have mass configurations similar to those mentioned above except possibly for a large baryon cloud with a central gap located near a black hole or some very massive sun or other massive object.

The major challenge for the neutron star slingshot is the avoidance of difficulties associated with a close approach. This seems possible due to the very short time spent close to the neutron star.

[178] The rate of spin of the uniship has to be acceptable from a human factors point of view.

17.2.1 Spinning Baryon Ring Gateway to the Megaverse

A spinning baryon ring generates a baryonic field that causes a moving baryon uniship in the "center" to spiral into the Megaverse. As soon as the uniship is in the Megaverse, outside our universe, it is no longer visible. Once in the Megaverse the uniship can use its own built-in propulsion system to travel to other universes.

One problem with this approach is the return to our universe. There is no spinning ring in other universes and the construction of one would be costly time-wise, and technology-wise, not to mention the large amount of mass that would have to be assembled to make a ring or cylinder. An alternative approach would be to use the slingshot mechanism for the return to our universe or the spinning uniship mechanism of section 4.2 below.

//

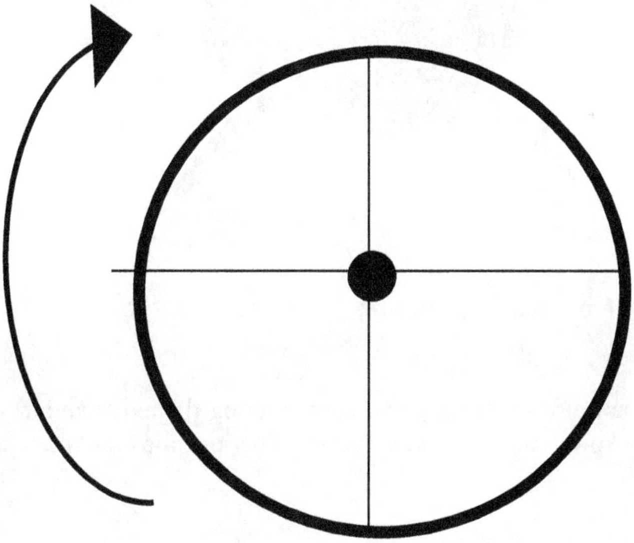

Figure 17.2. A clockwise rotating thick baryonic ring with a uniship at its center moving "straight out of the page."

17.2.2 Spinning Baryon Cylinder Gateway to the Megaverse

A uniship enters a baryonic cylinder and under the baryonic force begins to spiral from our universe into the Megaverse disappearing from view. It then can use its engines to traverse the Megaverse to other universes. It can also view other universes in the Megaverse shining with "baryonic light." The return to our universe would require a baryonic cylinder (or ring or spinning uniship (section 17.3 below) or slingshot maneuver) in the Megaverse to reenter our universe.

/

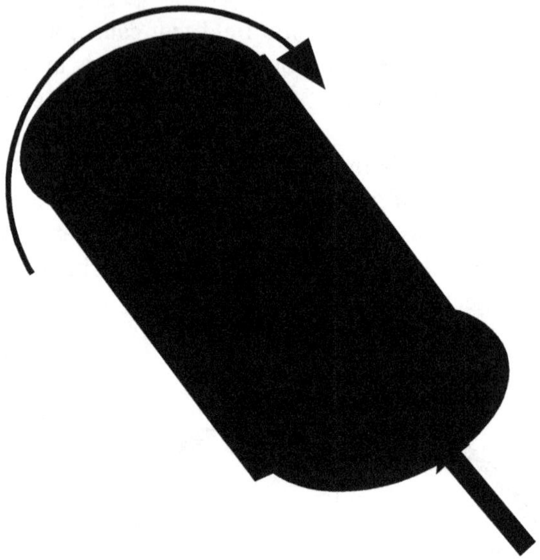

Figure 17.3. Depiction of a uniship entering a spinning baryonic cylinder along the cylinder's central axis.

Sending a uniship, with a large positive baryonic charge along the axis (center) of the cylinder will cause the uniship to spiral into the Megaverse. The uniship can then travel to other universes.

17.2.3 Rotating Baryon Mass Configuration Gateway to the Megaverse

One can also envision other baryon mass configurations that can be used to enable a uniship to escape from our universe into the Megaverse. One example of a mass configuration is four baryon spheres of large baryon number rotating uniformly. A uniship enters at high speed along their central axis and then spirals into the Megaverse by means of the baryonic force.

17.2.4 Disadvantages of Ring/Cylinder/Mass Configurations to Escape our Universe

The basic problem with all the possible above gateways is that they offer one way transit out of our universe and do not provide for a return. The approaches below provide a capability for a return to our universe.

All of these mechanisms require extremely large masses and are beyond our capabilities in the foreseeable future.

///
/

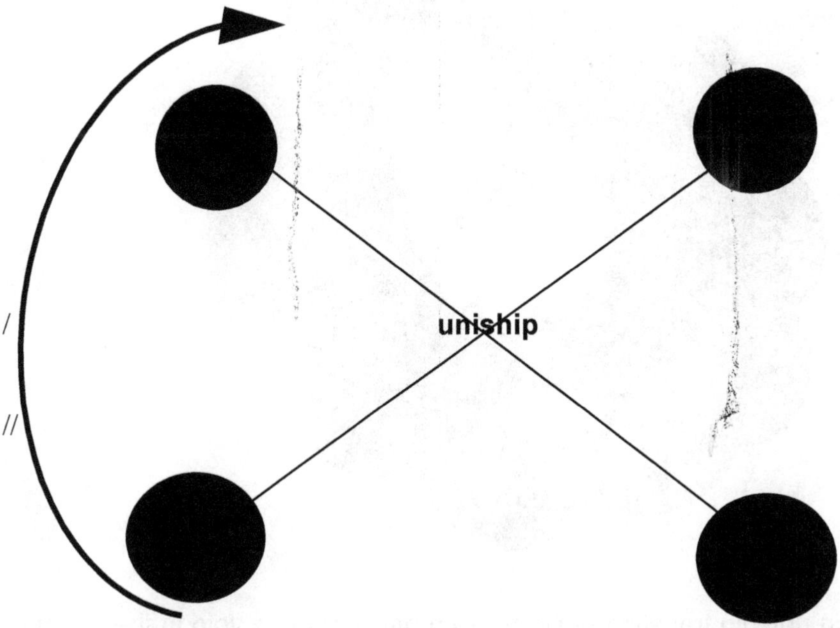

Figure 17.4. Depiction of a configuration of four large masses rotating around a common center. A uniship goes through the center ("out of the page") and enters the Megaverse.

This type of large masses configuration would be difficult to assemble and rotate in synchronization. However it might require less total mass than the ring or cylinder gateways.

17.3 Escape using a Rapidly Rotating Uniship

Above we explored using external large masses to cause a uniship to enter the Megaverse. In this section we consider a rotating uniship moving at high speed through a void in a massive cloud of baryons. It is necessary to move through a void to avoid the severe erosion that a high speed uniship would experience moving through a dense cloud of matter.

/

/

//

Figure 17.5. A rotating uniship traveling outward from page within a void in the midst of a massive baryon cloud.

As the fast rotating uniship moves through a void within the massive cloud of baryons it will experience a baryonic force that will cause it to spiral out of the universe into the surrounding Megaverse. A rotating uniship in a relatively static baryon cloud is equivalent to a non-rotating uniship within a rotating ring or cylinder or mass configuration.

This uniship escape mechanism could presumably be used in any large universe and could also be used to reenter our universe. It avoids the need to create gigantic mass configurations relying on Nature to provide massive clouds within a universe.

The major problem of the rotating uniship is the centripetal force that the crew would experience.

18. Uniship Design Topics

This chapter discusses some uniship design topics. It appeared in Blaha (2014c).

18.1 How Many Dimensions Exist Inside an Escaped Uniship?

When a uniship escapes from a universe into the Megaverse the questions arise, What is the dimensionality of the space within the uniship? How do the uniship occupants and equipment interface with the (D – 1) spatial dimensions outside the uniship?

The first issue that must be considered in answering those questions is the physical reality of the Megaverse. Are its coordinates a mathematical fiction? Or are they physically real? The answer seems to lie in the existence of a quantum field that depends directly on the D dimensions of the Megaverse. The existence of such a field gives Megaverse coordinates a physical reality since the quantum field, being a physical construct, could not be defined without the physical reality of the Megaverse. An additional requirement for a physical Megaverse is an interaction between the quantum field and the physical masses of our experience. The baryon gauge field $B_{ui}(y)$ that we have postulated previously fulfills both of these requirements. (See Blaha (2014) and earlier parts of this volume.) Thus the Megaverse which is the "ocean" of the set of "island universes" is a physically real entity.

The second issue is the relation of Megaverse coordinates and the coordinates of a universe. We note that every point in a universe's coordinates has a corresponding point in the Megaverse due to the relation:

$$y_i = f_i(x) \tag{18.1}$$

Thus every point in a universe has two sets of coordinates. This also holds also for the interior of a uniship.

In the uniship the occupants will live in a 4-dimensional flat space "bubble"[179] within the (D – 1)-dimensional Megaverse. The second question we raised concerned how the uniship occupants could interface with (D – 1)-dimensional space within and without its confines. This is considered in the next section.

---

[179] An analogous situation is a liquid sphere in zero gravity that splits into two spheres. Both spheres retain the same topology and other characteristics.

18.2 How Does a Uniship Control Direction and Maneuver in the D-dimensional Megaverse?

Transiting from a subspace with one set of dimensions to a larger set of dimensions in a larger space is a physically challenging task. One cannot simply use Euclid's algorithm for proving three spatial dimensions. For it assumes that we can move in the direction of any physical dimension. But if the available physical force to move in a given direction is not present then the dimension is physically inaccessible.

In the present case we enable a uniship to escape from our universe using, for example, the slingshot around a neutron star mechanism. After entering the Megaverse it will have some momentum in Megaverse directions beyond the three spatial dimensions of our universe. But that is not enough. A uniship must have the capability to move in any or all D – 1 directions in the Megaverse.

Thus we are led to propose a uniship designed having D – 1thrust exhausts with each direction of thrust pointing in a direction in (D – 1)-space. Thrust would be generated in a central chamber and then directed, using (perhaps) bending magnets, through one or more exhaust ports accelerating the uniship in a specified direction. In our universe all D - 1 exhaust ports and the direction mechanism would point in a variety of directions in our 3-space. As a uniship slingshots around the neutron star it will experience strong gravitational and baryonic tidal forces. With a correct alignment of the D - 1 exhaust ports and central chamber in three dimensions, the exhaust ports and chamber will open like an umbrella in all the directions of the (D – 1) Megaverse dimensions. Then the uniship will be able to accelerate in any direction in the Megaverse. Fig. 18.1 depicts the uniship before and after the slingshot maneuver.

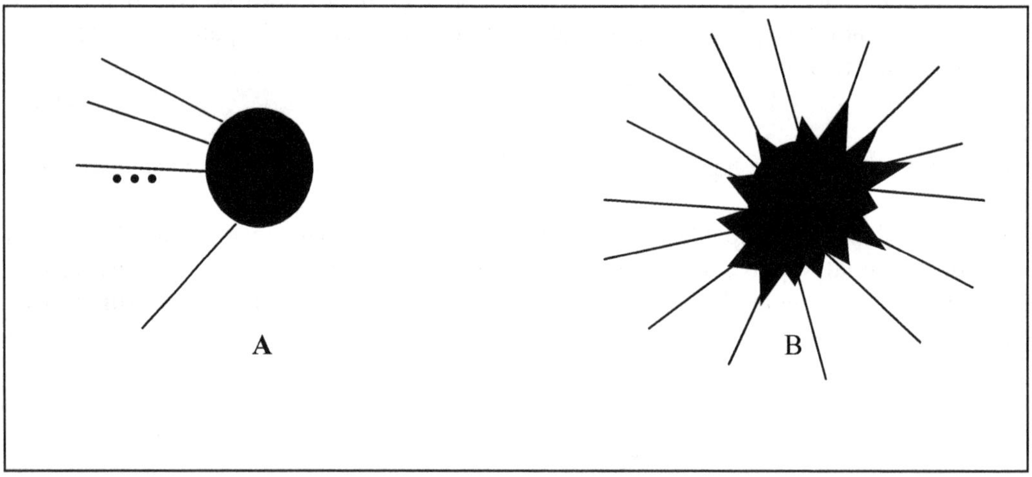

Figure 18.1. Symbolic depiction of the uniship thrust exhaust ports before (A) in 3 dimensions, and in (D – 1) dimensions after (B) the slingshot maneuver to exit our universe.

18.3 Power Source for Uniships

The distance scale of starship travel in our universe (and probably in other universes) between galaxies is of the order of millions of light years. Consequently starships will have to be powered by particle-antiparticle annihilation and/or fusion energy for longer trips, and nuclear energy, at minimum, for intra-galaxy trips.[180] Part of the power requirements are necessitated by the need for short travel times of a few years at most so that interstellar commerce, and migration, becomes feasible. For these reasons we wish to have starships capable of speeds up to tens of thousands times the speed of light.

The distance scale of travel in the Megaverse is likely to be trillions of light years. We anticipate universes are separated by trillions of light years. Due to these enormous distances, and the fast transportation time that we wish, uniships will require proton-antiproton annihilations on a large scale to power uniships. Proton-antiproton annihilation transforms mass entirely into energy unlike nuclear and fusion reactions where only a fraction of the masses of the particles are converted into energy yielding of the order of 1,000 the energy production of fusion.

18.4 Possible Uniship Design

We therefore require a large sphere containing protons that are densely packed. (An alternative that would avoid electromagnetic repulsion problems associated with densely packed protons would be a densely packed sphere of hydrogen atoms.) We also need a large sphere containing antiprotons that are densely packed. (Again an alternative that would avoid electromagnetic repulsion problems associated with densely packed antiprotons would be a densely packed sphere of anti-hydrogen atoms.) In addition to being the most powerful energy source available these spheres would form a baryon dipole that could play a role in the escape of a uniship from a universe. We therefore start with a uniship design of the form of Fig. 18.2.

Figure 18.2. An initial depiction of the overall design of a uniship. The upper and lower ovals contain a dense mass of protons and of antiprotons respectively. The antiproton sphere requires a confinement mechanism to avoid premature particle annihilation. (The ovals could alternatively contain liquid hydrogen and anti-hydrogen to avoid electromagnetic repulsion issues.) The module in front contains the crew quarters, equipment and cargo. The tail section contains a central region that controls the

[180] Blaha (2013a) and earlier books.

particle-anti-particle combustion process directing the thrust through one or more of the (D – 1) thrust chambers to power the uniship in the desired direction in the Megaverse.

18.5 Intra-Universe Starship Designs Summary

Starship design has a number of requirements to satisfy for trips within the galaxy and a more stringent set of requirements for travel to other galaxies in our universe. Uniships have a yet stricter set of requirements for travel to other universes within the Megaverse. The discussion in this section sets the background, based on our starship designs, for the consideration of important design details for uniships presented in succeeding chapters. Most of this section is abstracted from Blaha (2013a) and earlier books on nuclear ships, space guns, and starships. Appendix A of Blaha (2013a) briefly overviews the designs in these earlier books for mass transit from the earth to space, travel in the solar system, and, most importantly, starship travel within our galaxy and to other galaxies.

18.5.1 Types of Starships

There are two general types of starships due to the nature of starship drives: starships based on circular accelerators and starships based on linear accelerators. In both cases we assume the process that creates the quark-gluon plasma thrust is a stream of collisions of spherules of some material.

18.5.2 Starship Engine Energy Sources

The starship engine that we have designed requires massive amounts of energy. At this point in time the only feasible energy source for starships is nuclear energy. It is reasonable to expect that fusion energy, a more concentrated energy source, will become a reality within the next thirty years. In either case the starship will need an energy source to drive the spherule accelerator rings and associated devices of the engine for periods up to perhaps a few months or years, then turn off for perhaps many years, and then resume operations for further maneuvers.

In the extreme case of travel to another galaxy, the energy source will need to turn off for up to millions of years of starship time. While the energy source is turned off, a residual "battery" will need to operate to support monitoring the progress of time, activating the startup of the main energy source, and possibly to detect and monitor objects ahead of the starship in the line of flight. This battery source may well be a plutonium (or longer lived) source similar to those used in current space probes.[181]

The main energy source, if it is a nuclear reactor of some kind, will probably have to be a reactor that is different from current nuclear reactors. Chapter 5 of Blaha (2013a) describes long shelf life nuclear reactors that could meet this need. Since the startup process from a

[181] We note that a natural nuclear reactor existed in the Congo Region of Central Africa for millions of years. (Parenthetical note: Could this be the stimulus for the rapid evolution of species in Africa including very early Mankind?)

battery driven state needs to be gradual due to a "small" battery, it appears the set of nuclear reactors would be composed of perhaps five reactors of increasing size. The battery starts the smallest reactor by concentrating its nuclear fuel. The smallest reactor then generates the energy to concentrate that fuel, and start the second smallest reactor, and so on until the main reactor(s) is started. At this point the accelerators power up and starship thrust begins.

If the source of the energy is fusion energy then the startup process might begin in small stages in the boot up of the fusion reaction through fusing larger and larger amounts of (perhaps) ^{3}He with increasingly powerful laser beams a la the tokamak approach.

During the coasting period of a starship, nuclear reactors should be powered down to conserve nuclear fuel. When powering down a nuclear reactor power source the nuclear material (U^{235} or plutonium) (the reactor fuel) residing in the medium of the reactors would be diluted to sharply reduce fission reactions to "near zero" using the energy of the next largest reactor. The smallest reactor would be powered down by a battery. This battery would retain enough energy to bring the smallest reactor back up after the coasting period ends. Then the reactors would boot up in turn to provide energy to the vehicle. (The battery would be at extremely low temperature during a coasting phase and thus not lose a significant amount of electrical power.)

In the case of a fusion power source a battery could be used to initiate the fusion power. A gradual turnoff process could execute at the start of a coasting phase to bring the fusion process to zero in such a way that the battery could initiate the boot up process for the fusion power source at the end of a coasting period.

18.5.3 An Alternate Starship Accelerator Ring Engine Design

In an earlier work Blaha (2009b) we proposed an alternate accelerator ring design for quark-gluon fluid acceleration. It appears that this design is not feasible with current or near term (one hundred years) technology. The reasons are:

1. The quark-gluon fluid ring is not feasible because fireballs cannot be injected and accelerated to form a ring in a few fm/c - the time available.

2. A quark-gluon fluid ring would be similar to fusion tokamaks but much more challenging in its requirements for confinement and stability due to the much higher density and temperature of a quark-gluon fluid ring.

18.5.4 Some Starship Configurations

There are a variety of possible configurations for faster than light starships. In this subsection we categorize the starship configurations by their thrusters.

18.5.4.1 Rear Thrust Exhausts on Circular Accelerator Starships

In Blaha (2010a), and later books, quarks and gluons were shown to have complex 3-momenta. The real part of the 3-momenta of each quark was orthogonal to the imaginary part of its 3-momenta.

If we "add" pairs of exiting quarks to create a complex 3-momentum to the rear, then a complex total pair 3-momentum is created that generates a rearward thrust. The complex thrust can be used in a one thrust exhausts, or two thrust (or multi-thrust) exhausts as shown in Figs. 18.3 and 18.4.

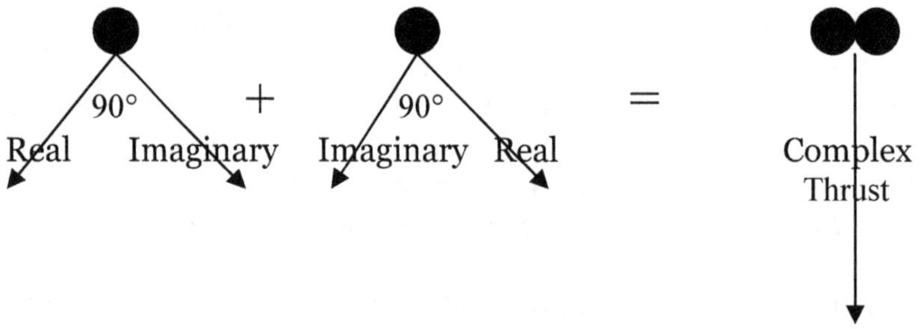

Figure 18.3. A circular accelerator starship with one thrust exhaust.

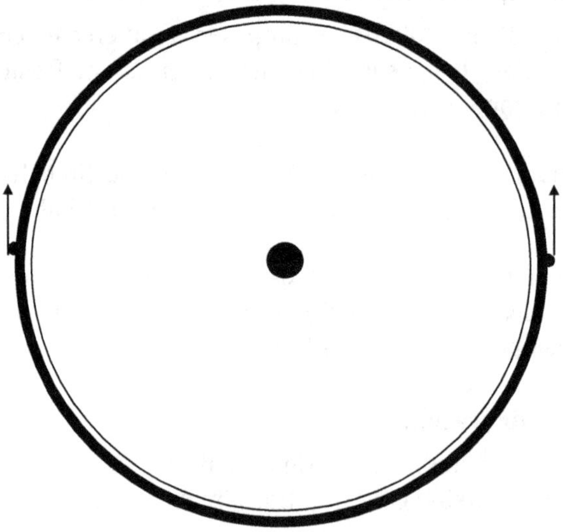

Figure 18.4. Top view of a circular starship with two complex thrust exhausts.

These types of starships generate thrust that only moves the starship forward in space. They do not cause the starship to rotate. The next subsection introduces the use of the imaginary perpendicular part of a starship's thrust to cause the starship to rotate.

18.5.4.2 Rotating Circular and Cylindrical Starships due to Imaginary Part of Thrust

In the case of disc-shaped and cylindrical starships the imaginary part of the quark-gluon thrust can be directed by magnets to be tangent to the circular edge of the starship. The resulting tangential force causes the starship to rotate through an imaginary angle. The rotation creates a *negative* centripetal force in the starship – "artificial gravity" that varies with the distance from the central axis of the starship. Fig. 18.5 shows a "top view" of a starship.

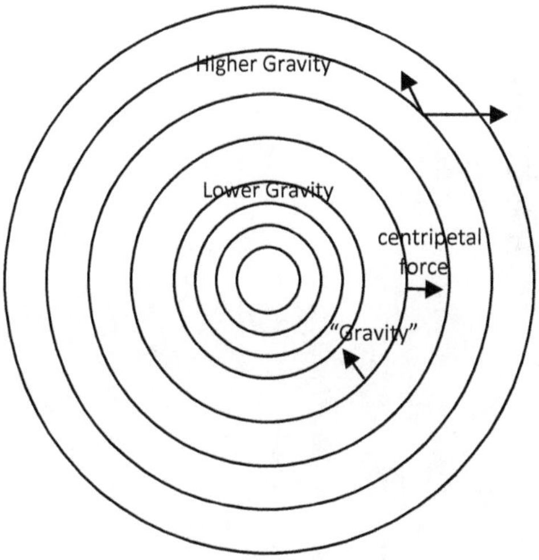

Figure 18.5. Top View of a Disc-like or Cylindrical Starship. View of cargo/people levels of inner hub. The center has lower "gravity." Circles indicate levels of equal artificial gravity in a disc-shaped or cylindrical starship. The outer parts have higher "gravity." The "gravity" force is inward towards the center on all levels. The centripetal force is outward from the center as shown by the arrow in the diagram and the discussion below (opposite to the direction of conventional centripetal force.) The artificial "gravity" force is thus inward to the center.

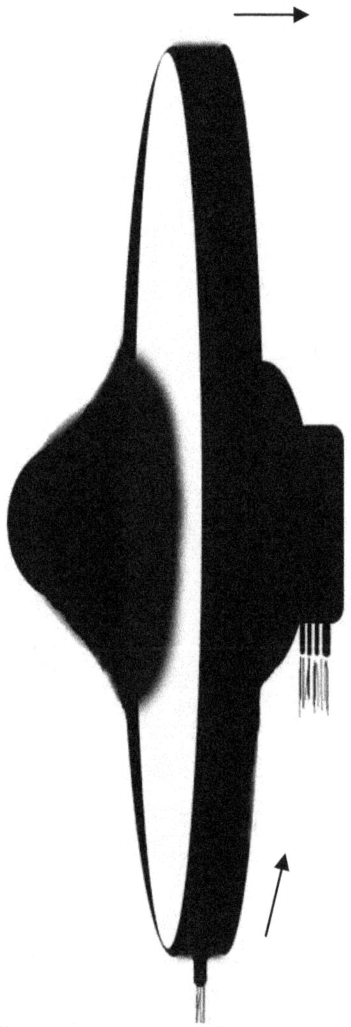

Figure 18.6. A disc-like starship with thrust consisting of a real and imaginary part. The real part drives the ship to the left. The imaginary part is tangent to the edge of the starship causing it to rotate counterclockwise.

Fig. 18.6 is an example of a disc-shaped starship with the thrust direction displayed. The horizontal thrust arrows represent the real part of the quark-gluon thrust. It accelerates the starship to the left in the figure. The vertical arrows represent the imaginary part of the thrust. It rotates the starship in a counterclockwise direction.

18.5.4.3 Artificial Gravity Levels - Rotating Circular and Cylindrical Starships

In rotating starships the real part of the thrust propels a starship. The combination of the real and imaginary parts of the thrust enables a starship to exceed the speed of light. The imaginary part of the thrust causes the starship to spin around its central axis.

We will examine the case of a cylindrical starship and see how the "artificial gravity" emerges as a result of the imaginary part of the thrust. Fig. 18.7 shows a cylindrical starship and Fig. 18.5 shows the cylindrical coordinate system used to calculate its motion.

The force generated by the quark-gluon thrust has the form

$$\mathbf{F} = g_r\check{\mathbf{z}} + ig_i\boldsymbol{\o} \qquad (18.2)$$

in the starship's cylindrical coordinates rest frame where $\check{z}$ is a unit vector in the positive z direction and **ø** is a unit vector in the positive ø angle direction. g_r and g_i are constants specifying the thrust. g_r is the starship's mass times its real-valued acceleration in the $\check{z}$ direction. g_i is the starship's mass times its imaginary-valued acceleration in the **ø** direction. The time derivative of the momentum **p** of a small part of mass m_0 at the edge of the starship is

$$d\mathbf{p}/dt = m_0(\gamma z')'\check{\mathbf{z}} + m_0[(\gamma\rho\o')' + \gamma\rho'\o')]\boldsymbol{\o} + m_0[-\gamma\rho(\o')^2 + (\gamma\rho')']\boldsymbol{\rho} \qquad (18.3)$$

/where **ρ** is the unit vector in the positive ρ direction and the ′ symbol indicates a time derivative.

Figure 18.7. Cigar shaped starship with “horizontal accelerator ring(s). The real part of the thrust points downward. The imaginary part of the thrust is horizontal and tangent to the cigar surface. The fins are for supplementary nuclear maneuvering rockets.

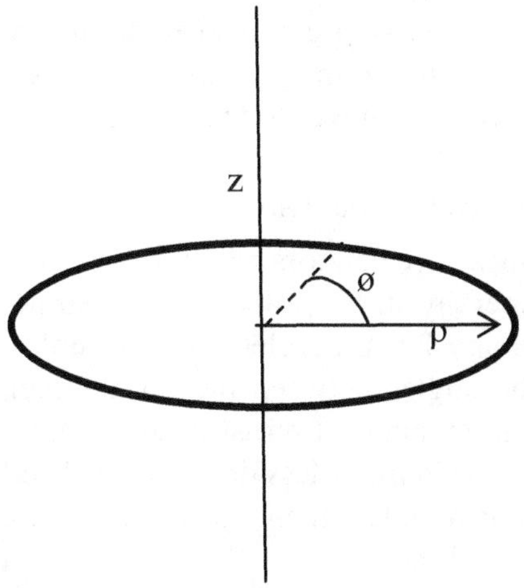

Figure 18.8. Starship cylindrical coordinate system. The diamond-shaped Starship in Fig. 18.9 is centered on the z axis.

If we set $\rho' = 0$ since there is no force component in the **ρ** direction and the starship is not contracting, then setting

$$\mathbf{F} = d\mathbf{p}/dt \tag{18.4}$$

implies

$$m_0\rho(\gamma ø')' = ig_i \tag{18.5}$$

The centripetal force is

$$F_{centripetal} = m_0\gamma v_c^2/\rho \tag{18.6}$$

where v_c is the rotational velocity of the mass. It satisfies

$$v_c = \rho ø' \tag{18.7}$$

Since $ø'$ is imaginary by eq. 18.5, v_c Is imaginary and thus the centripetal force $F_{centripetal}$ is negative by eq. 18.6. The centripetal force is "opposite" to the centripetal force normally experienced – away from the center. The artificial "gravity" force – really the opposite of the centripetal force – is towards the center. See Fig. 18.7.

The artificial "gravity" engendered by the rotation raises an issue. The acceleration of the rotation implied by eqns. 18.3 and 18.4, if continued, would generate an enormous inward "gravitational" force crushing the starship's occupants. This problem can be solved by causing an oscillation in the rotation between clockwise and counter-clockwise directions by repeatedly flipping the value of the imaginary force g_i (eq. 18.2) to maintain a constant rotation speed (and thus have constant values of "gravity.")

A circular (disc-shaped) starship can also rotate to create artificial gravity. Fig. 18.6 illustrates a rotating disc starship, Again the gravitational force is towards the center with lower gravity in the central region. The disc moves to the left due to the real thrust.

18.5.4.4 Starships Based on Linear Particle Accelerators

Starships based on linear accelerators can have a variety of forms. Fig. 18.9 contains a diamond-shaped starship. Generally starships based on linear accelerators need great length – of the order of miles for the primary linear accelerator tubes unless a very rapid linear accelerator mechanism is developed. The angles between the linear accelerator tubes are determined by maximizing the efficiency, and amount, of thrust generation. Thus the shape of the *lower half* of the diamond-shaped starship is more or less determined. The lower part of the starship can be sharp edged diamond shaped or rounded diamond shaped. The upper part of the starship should minimize the effects of space dust.

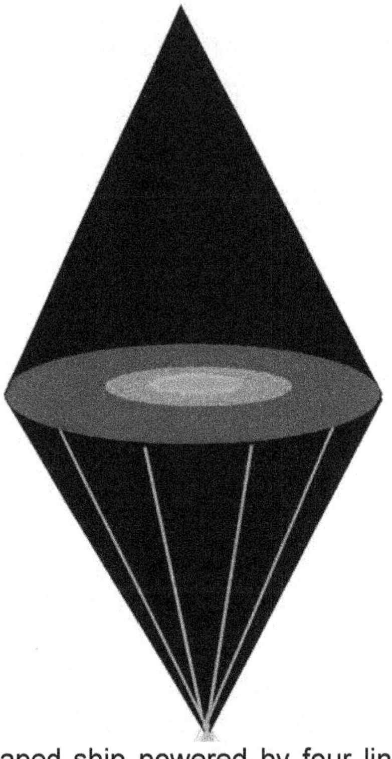

Figure 18.9. A diamond shaped ship powered by four linear accelerators. The lower part has the accelerators, magnets, nuclear reactors, and propellant. The upper part contains the crew, cargo, shielding and nuclear shuttles for exploring a solar system.

19. Megaverse Communications

Quantum communication is a rapidly developing field. At this point in time it has only been demonstrated for short ranges. Eventually it should mature as a long distance communication device such as that described in Blaha (2014c). Its great point is its instantaneous nature. Instantaneous communication at very great distances becomes feasible.

Thus one can hope to communicate across the universe. One can also hope for instantaneous communication with ships in the Megaverse. Sending a uniship into the Megaverse should not interfere with quantum communication using a detector at home and another that traversed into the Megaverse on a uniship. The difference in the number of dimensions should not affect the communication between the quantum devices. We expect that quantum states straddling the universe-Megaverse border will not go "out of sync."

Therefore quantum communication offers the possibility of instantaneous communications in our universe, and with uniships/colonies in the Megaverse and other Megaverse universes.

19.1 Rapid Interstellar Communication

Recent work on quantum entanglement suggests that instantaneous communication may be possible using this mechanism if an advanced long range form of this laboratory phenomenon can be developed. Quantum entanglement can transcend the borders of universes since it is based on coordinated parts of a quantum state. Its instantaneous nature, which has been verified to great accuracy in recent experiments, makes it the ideal mechanism for communication over trillions of light years. We will describe the application of this concept in more detail in the following sections.

19.2 "Instantaneous" Interstellar Communication

Once a uniship capability is achieved it will clearly necessitate a very rapid, if not instantaneous, means of communication. All electromagnetic means of communication are limited by the speed of light and are thus insufficient for multi-million light year communication. If neutrinos are tachyons (faster than light) then they could provide a communications channel except that neutrino detection is very difficult, not reliable, and would require massive detectors that would be an unacceptable addition to the mass of a uniship. More importantly, because neutrinos are extremely light particles their speed is not much more than the speed of light at best.

The only possible method appears to be a quantum entanglement mechanism – currently a subject of intense scientific interest. Based on current thinking about this form of quantum communication it will have the following very desirable features:

1. It is a 1:1 form of communication with no possibility of being intercepted by others.
2. It requires a small amount of power no matter what the distance.
3. It is instantaneous and thus gives direct real time communications over any distance – even trillions of light years.

If history is any guide, the development of inter-universe communications will be similar to the development of telecommunications over the past 150 years, but on a much longer development time scale. Thus we anticipate that it will begin with a primitive Morse code equivalent, and progress eventually to fast digital transmission of images and data. We anticipate bilateral switchboards initially that eventually will lead to communications with uniships beyond our universe in the Megaverse or other universes. Obviously this capability would be needed for exploration – particularly by the initial robot-driven uniships, and for communications between colonies and earth scientific or commercial reasons in other universes.

The basic mechanism will consist of a bilateral quantum entanglement setup that begins as two electrons[182] of opposite spins in a quantum state with total spin zero. Each electron is nudged into a magnetic bottle that does not affect their joint spin state.[183] One bottle is retained on earth; the other bottle is placed in a uniship. As the uniship travels the state of the electron spin within its bottle can be periodically sampled but without changing its state.[184] This can also be done on the earth based electron in its bottle. If either electron's spin is flipped the spin of the other electron flips instantaneously no matter what the distance. Thus instantaneous communication of one computer bit takes place.

Eight such bottle pairs allow us by flipping bits to exchange bytes of data. Because of the time contraction associated with much faster than light uniships the byte change must be almost instantaneous for effective communication between a uniship and earth. This fast exchange can be done by ultrafast computers.

Eventually arrays of "bottles" can transmit bytes in bulk in support of large data and image transfer. One can envision electronic switchboards eventually linking arrays of bottles to form a network with a set of uniships and/or colonies. The thought processes and designs are similar to those used in telecommunications.

---

[182] Protons would be another reasonable alternative.

[183] Several experimental groups have recently been able to detect parts within quantum states without affecting the overall quantum state.

[184] 2012 Nobel Prize winner Serge Haroche of France developed ways of detecting the state of particles without disturbing their quantum state.

It is important to note that quantum communications does not require powerful transmitters Thus quantum communications is energy efficient.

19.3 Experimental Support for Instantaneous Quantum Entanglement Data Transfer

One might ask if instantaneous quantum data transfer is possible. Both quantum theory and numerous experiments have shown that instantaneous data transfer via entangled pairs works at large distances.[185]

19.4 Interstellar Communications and SETI

If our (and others) suggestion that quantum communication is the only reasonable way for communications at large distances, then this might be the reason for SETI's failure to find communications by alien civilizations. Aliens may very well not be communicating by radio or laser waves.

It is important to note that quantum communication, as we have proposed it, is inherently private 1:1 communication with no visible manifestations for others to detect of which we are aware.

19.5 Quantum Entanglement and the Megaverse

In 1935 Einstein, Podolsky, and Rosen,[186] and shortly afterwards Erwin Schrödinger, began the discussion of what has become known as *quantum entanglement*. It has become a subject of growing, widespread interest since its validity has been demonstrated in numerous experiments.

Quantum entanglement can be described as a phenomenon in which a quantum state of two or more particles evolves to a point where the particles are separated by a distance such that there is no (hitherto) known mechanism by which the particles can communicate. The particles are separated by a space-like distance and thus cannot classically "communicate" at speeds at or below the speed of light. Yet a measurement of a property of one of the particles causes an instantaneous change in the other particles initially entangled with it.

Einstein's unhappiness with quantum entanglement is succinctly expressed with his often quoted description of quantum entanglement, "spooky action at a distance." Yet, since the 1930's, experiments have repeatedly shown that quantum entanglement is correct.

The bothersome issue which many physicists, starting with Einstein, are concerned is the mechanism by which instantaneous quantum effects happen. Dynamically we know that the evolution of a quantum system in the usual coordinates of our universe (whether a flat or curved space-time) yields quantum entanglement. However, a quantum measurement is outside the

---

[185] Matson, John, Quantum Teleportation Achieved Over Record Distances, Nature, 13 August 2012.

[186] Einstein A, Podolsky B, and Rosen N, "Can Quantum-Mechanical Description of Physical Reality Be Considered Complete?", Phys. Rev. **47**, 777 (1935).

dynamical equations being a sharp transition between the quantum state before and after the measurement – only regulated by continuity conditions between the initial quantum state and the post-measurement quantum state.

In the case of a two particle system in an initial quantum state the invariant distance (squared) between the parts of the system when they are widely separated is

$$\Delta s^2 = \Delta t^2 - \Delta \mathbf{x}^2 \tag{19.1}$$

with $\Delta t = 0$ at equal time. The coordinates are real-valued and governed by the transformations of the Lorentz group in the usual discussions of quantum entanglement. Thus the quandary of instantaneous changes in the parts of a system separated by a space-like distance.

We now suggest that quantum measurements should be viewed within the framework of complex-valued 4-dimensional space-time and the complex Lorentz group which we have shown in prior books leads to the form of The Standard Model of Elementary Particles (and thus quantum theory).[187]

If complex coordinates are the proper coordinates for observers then quantum entanglement is understandable. If a state is created and then parts of the state separate, measurement of part of the state at a certain time can set the other part of the state instantaneously. The invariant distance between the parts are space-like separated in real-valued space-time at equal time:

$$\Delta s^2 = -\Delta \mathbf{x}^2 = -(\Delta x^2 + \Delta y^2 + \Delta z^2) < 0 \tag{19.2}$$

But they can be transformed by a complex Lorentz group transformation $R^i_{\ j}$

$$[\Delta \mathbf{x}']^i = R^i_{\ j}\, [\Delta \mathbf{x}]^j \tag{19.3}$$

to zero spatial distance giving

$$\Delta s'^2 = -\Delta \mathbf{x}'^2 = 0 \tag{19.4}$$

In this new coordinate system the time difference is not zero – it is a pure imaginary number $\Delta t' = i|\Delta t'|$ thus maintaining the invariance of the full invariant interval:

$$\Delta s^2 = \Delta t^2 - \Delta \mathbf{x}^2 = \Delta s'^2 = \Delta t'^2 - \Delta \mathbf{x}'^2 \tag{19.5}$$

---

[187] We also showed that the Wheeler-DeWitt equation of quantum gravity when generalized to complex-valued General Relativity explains many of the features seen in the universe such as the web of clusters of galaxies and the lopsided nature of the universe. Thus we conclude that the Megaverse and its universes have complex-valued coordinates. Some relevant references are Blaha (2011c), (2012a), (2012b), (2013a), (2013b), and (2014a). Much of the content of these books were introduced in earlier books by the author.

Thus, using *complex* coordinates,[188] which are truly the fundamental coordinates of space-time, all spatial distances can be mapped to zero and the apparent separation suggesting quantum entanglement at a distance is a result of the map of complex-valued coordinates to the real-valued coordinates. The complex coordinates of our universe are related to the coordinates of the complex D-dimensional Megaverse by

$$y_i = f_i(x) \tag{19.6}$$

for i = 1, … , D where x is a complex 4-vector in our universe and the y_i are the corresponding coordinates in the Megaverse. The Quantum observer can make a measurement in either set of coordinates in a reference frame where the spatial location of the apparently separated parts of the quantum state coincide.

Spatial separations at equal time can thus be mapped to zero for observations (measurements) in complex coordinates while the dynamical evolution of the quantum state is expressed in real-valued coordinates. These real-valued coordinates can be related to the complex coordinates of our universe or the Megaverse.

We conclude that there is no mysterious "spooky action at a distance" in quantum entanglement. The mystery is explained if we use the right coordinate system for the observer: complex-valued coordinates.

Thus all parts of a quantum state can be viewed as being located at the same spatial point in the right reference frame. Properly viewed, there is no mystery to quantum entanglement.

Returning to our subject: Despite the disappearance of mystery in quantum entanglement, it still remains an exciting topic in view of its potential applications such as communicating instantaneously at interstellar, intergalactic, and Megaverse distances which we have discussed elsewhere.

19.6 An Improbable Megaverse Connection to Spirits, the "Spirit World", and UFOs

Ghosts, spirits, and spiritual phenomena have been associated with "the fourth dimension" and other extra dimensions since the 1850s. UFOs are also often described as coming from or through other dimensions. Some scientists have also associated physical phenomena with other dimensions.

We do not believe these characterizations of phenomena as artifacts of other dimensions to be correct in the manner in which they are described.

[188] As shown in our earlier books complex-valued coordinates in any coordinate system can be mapped into the real-valued physical coordinates that we experience using the Reality group. See Blaha (2011c) and other books by the author for more details.

However, we do think that one can simulate spiritual-like phenomena and UFO-like phenomena using the Megaverse. This section will briefly outline these possibilities. The hope is to forestall attempts to use our Megaverse theory to bolster support for these phenomena.

19.6.1 "Spiritual" Phenomena and the Megaverse

Spiritual phenomena have several types: the appearance of visions of people or things of one sort or another; material objects passing through solid objects; unseen voices; and so on. All of these phenomena could be simulated using the Megaverse to make things appear from the Megaverse or to go from within the universe through the Megaverse and reappear in our universe again.

Some examples on Megaverse simulations are:

1. Using a "Sidestep" into the Megaverse to Circumvent Solid Obstacles in our Universe.

2. Going through a Solid One Dimensional Wall. A 1-dimensional example is

------------------|----------------

3. Persons or things going through Solid Three Dimensional Walls.

While these "partly in and partly out of the universe" phenomena can be simulated by a Megaverse agent, the agent's nature and purpose are not determinable from physical considerations of the Megaverse.

19.6.2 UFO Phenomena and the Megaverse

UFOs have been "seen" in the skies in many regions of the earth. They are often characterized as having high speeds, extremely high accelerations, and the ability to make abrupt changes in direction at high speeds. All of these phenomena are understandable if we are seeing objects in the Megaverse where modest changes in speed, acceleration and direction in Megaverse coordinates can map to the UFO movements that we see in our universe's coordinates.

Beyond stating these phenomena can be understood from a Megaverse perspective we can say no more. They may be real. Their purpose, if they are real, cannot be discerned. Megaverse physics cannot enlighten us on these subjects.

/

20. Traveling and Navigating in the Megaverse

I saw Eternity the other night,
Like a great ring of pure and endless light,
All calm as it was bright;
"The World" – Henry Vaughan

Upon entering the Megaverse the first uniship will want to scan for universes and simply create charts of "nearby" interesting universes rather like astronomers on earth began by charting the galaxies in the heavens and looking for novel features and phenomena. Subsequently, the first uniship, and the following uniships, will travel to universes of interest and begin their exploration. What they will encounter is anyone's guess. But we have good reasons to believe that other universes will have similar physical laws although they may have different values for physical constants such as elementary particle masses and coupling constants. The reasons are:

1. We believe the dimensionality of universes is fixed by the principles of Asynchronous Logic that make four space-time dimensions the minimal acceptable number of dimensions. (We note that higher dimensional space-times are not excluded by these principles. But Nature tends to favor extremums – a minimal number of dimensions in the present case. And Nature also tends to repeat successful designs.)

2. We have shown that complex four dimensional space-time leads to The Standard Model directly (particularly the form of the fermion mass spectrum) with an additional $SU(2)\otimes U(1)$ that we associate with the Dark Matter sector.

3. Since all space and time measurements yield real-valued numbers we require a Reality group to map complex space-time coordinates to real-valued coordinates. The Reality group gives part of the group structure of an extended Standard Model $SU(3)\otimes SU(2)\otimes U(1)\otimes SU(2)\otimes U(1)$. Thus the form of the Standard Model (somewhat extended) is directly based on Logic and geometry.

4. Gravitation is also based on geometry.

Thus we have the form of physical theory based on geometry although the numerical constants in the theory remain to be specified.

We conclude almost every universe has a similar fundamental physics theory possibly with different physical constants. We say "almost every" because the possibility of irregular universes cannot be ruled out. Nature can choose the bizarre at times. In Blaha (2014) we describe the types of universes that can occur. In earlier books, such as those listed at the beginning of this book, we developed the geometric theory of Physics described briefly above.

We have described the necessary structure of universes realizing that they may contain unusual phenomena and may contain alien civilizations with which we can have scientific and cultural exchanges. Consequently it behooves us to explore the Megaverse to extend the range of human knowledge and possibly find reasons for establishing commerce.

In this chapter we will consider aspects of uniship travel, exploration, and navigation. In the following chapters 21 – 26 we will consider other aspects of uniships.

20.1 Uniship Range and Travel Times

Ideally we would like uniships capable of traveling large distances of the order of trillions of light years in relatively short times of the order of years or, at least, decades. These wishes cannot be accomplished with current or foreseeable technology. Crews can only withstand modest accelerations up to 8g at present and probably not much more with foreseeable space medicine advances. Thus accelerating to millions of times the speed of light will take a considerable amount of time.

Given the acceleration time requirements trips between universes become lengthy – perhaps decades of earth years. On a uniship, at these very high speeds the occupants and equipment will see time progress rapidly – at a rate equal to the speed measured in units of c (the speed of light) times earth time because of relativitistic dilation. A uniship traveling at 1,000,000c will experience time increasing at 1,000,000 times earth time. An earth year thus becomes a million years of uniship time. This time dilation effect places stringent requirements on uniship equipment and requires suspended animation for the crew. We will discuss these issues in more detail later.

On the positive side, if the uniship makes a roundtrip at that speed only two earth years will have elapsed, and if they were in perfect suspended animation, the crew's body clocks will have aged only two years in sync with earth time.

The preceding discussion has ignored one major point – the energy required to accelerate to millions of c is enormous. That is why we have to use the most powerful "concentrated" energy available. We hope that energy derived in quantity from particle annihilation will be available 50,000 years hence if not much sooner. Otherwise travel into the Megaverse becomes analogous to an ant swimming across the Pacific Ocean.

The above comments cannot be viewed as encouraging. But with scientific and technological advances in the next 50,000 years there is hope. And there is good reason to begin thinking of that ultimate goal and how we may eventually reach it.

20.2 Uniship Baryonic D-Dimensional Observation/Seeing Techniques

If a uniship enters the Megaverse of the Megaverse there should be countless universes in view if there is a method for seeing them. However the ability to detect universes is very limited. We cannot detect gravitational waves from universes because their gravity is confined to their universes.

More importantly we cannot detect electromagnetic waves emerging from universes because they are confined to the universe within which they were created.

Consequently, the radiation from univereses that we can expect to encounter in the Megaverse is baryonic or electromagnetic radiation. There are no other known long range types of radiation emitted by universes. This creates a quandary that we can only begin to address at our present state of knowledge. Observing baryonic radiation will be a difficult task – not just because of the weakness of the baryonic field coupling constant but also for several important reasons:

1. We can't see baryonic radiation either visually or through technology at present. No detectors.

2. We cannot focus on the source(s) of baryonic radiation so as to distinguish their direction and distribution.

3. We have no mechanism to magnify the pattern of incoming baryonic radiation. No lenses, telescopes or other viewing mechanisms with "zoom" capabilities. And no technology to amplify baryonic radiation.

4. Baryonic radiation is (D – 1)-dimensional. We would have to be able to create 3-dimensional hologram projections that provide an intelligible view. The holographic images would have to be manipulated to enable navigation.

5. Baryonic radiation has (D – 2) polarizations which would provide information on the nature of universes. The detection and analysis of these polarizations is currently beyond our capabilities.

Our inability to detect gravity waves (except in recent studies of the Big Bang – BICEP2) shows the difficulty of detecting and analyzing baryonic radiation.

20.2.1 Relativistic Effects on Radiation

The baryonic view of the Megaverse that a uniship crew "sees", when the uniship is traveling faster than the speed of light, is very different from its view when traveling at low speeds of a few tens of miles per second.

An observer on a uniship traveling at a relativistic speed near, but below, the speed of light will detect universes with baryonic or electromagnetic radiation compressed to within a cone in the frontal direction of the uniship (Fig. 20.1). The baryonic radiation cone becomes narrower as the speed of light is approached due to aberration and in the limit as the speed approaches the speed of
light becomes a point directly ahead of the uniship.

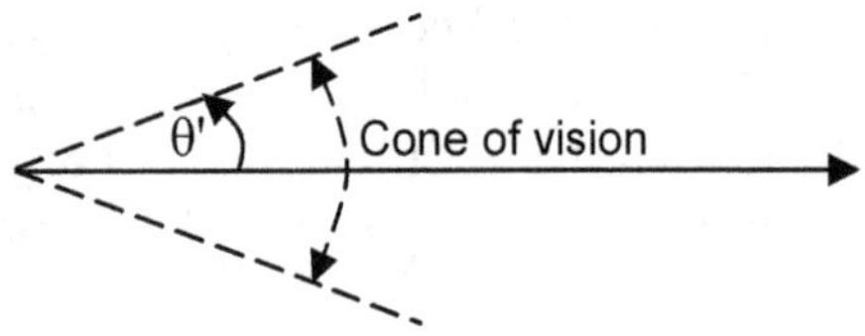

Figure 20.1. Baryonic (and electromagnetic) radiation cone of visibility around direction of uniship motion in the uniship coordinate system with the angle θ' determined by eq. 20.1 for sublight uniship speeds.

The relativistic equation for baryonic radiation aberration is

$$\cos \theta' = (\cos \theta + \beta)/(1 + \beta \cos \theta) \tag{20.1}$$

where θ is the angle of a universe relative to the uniship's direction of motion as measured in the Megaverse coordinate system and θ' is the angle of a universe relative to the uniship's direction of motion as measured in the uniship's coordinate system.

The inverse relation is

$$\cos \theta = (\cos \theta' - \beta)/(1 - \beta \cos \theta') \tag{20.2}$$

20.2.1.1 Sublight Case: β < 1

As $\beta \rightarrow 1$ (the speed of light) eq. 20.1 indicates $\theta' \rightarrow 0°$ showing the entire view of the multiuniverse baryonic radiation is compressed to the forward direction. Fig. 20.1 shows the cone of visibility for a uniship traveling near the speed of light at perhaps .6c - .9c. The cone angle θ' satisfies

$$\cos \theta' > \beta \tag{20.3}$$

The rest of the field of view of the uniship is total baryonic radiation blackness except the point in the directly rearward direction (θ' = 180°) for any object at θ = 180°.

20.2.1.2 Superluminal Case: β > 1

For $\beta > 1$ eqns. 7.1 and 7.2 still hold and there is a cone of baryonic radiation visibility similar to that depicted in Fig. 20.1. However the cone angle θ' for superluminal speeds, $\beta > 1$, satisfies the relation

$$\cos \theta' > 1/\beta \qquad (20.4)$$

The rest of the field of view of the uniship is total baryonic radiation blackness, as in the sub-light speed case, except the point in the directly rearward direction ($\theta' = 180°$). We note that as β gets very large the cone of visibility becomes larger. At $\beta = \infty$ the cone of visibility becomes the angular region between $\theta' = 0°$ and $\theta' = 90°$ (the forward hemisphere).

20.2.1.3 Superluminal Uniship Visibility

As a result, "visual" baryonic radiation navigation at high superluminal speeds becomes difficult unless we develop an electronic imaging system that "undoes" the effects of aberration and enables "visual" baryonic radiation navigation.

A further problem is the location of a destination. If we send a uniship to a far universe we have to project the location of the universe at the time the uniship arrives based on the universe's current motion. If the motion of the universe is modified by forces exerted by nearby universes as baryonic radiation from the destination universe travels to the uniship, or if the universe's motion is not accurately determined, a uniship could arrive at a point that is still some distance from the destination universe. Thus navigation to a far universe is a significant issue.

20.2.1.4 Effect of Doppler Shift at Superluminal Speeds

A uniship traveling at relativistic sublight speeds will see universes having their baryonic radiation spectrum (color) changed significantly due to the Doppler Shift effect. At superluminal speeds the Doppler Shift will also change the "colors" of objects "seen" by the uniship.

This issue is again surmountable if we use electronic imaging techniques to "undo" the Doppler shift and thus display universes with their true baryonic radiation spectrum in the uniship's reference frame.

The relativistic Doppler shift for sublight speeds of a baryonic radiation wave of frequency ν is given by

$$\nu = \nu_0(1 - \beta^2)^{1/2}/(1 - \beta \cos \theta') \qquad (20.5)$$

where ν_0 is the frequency of the baryonic radiation emitted by the source universe and θ' is the angle of the source relative to the uniship's velocity (Fig. 20.2).

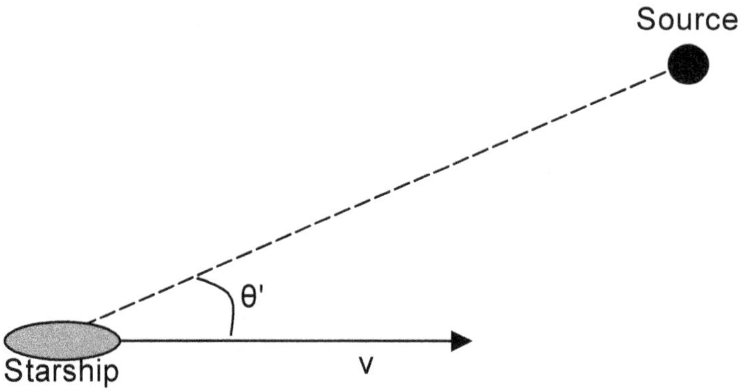

Figure 20.2. The angle of a universe θ' with respect to the uniship's velocity v.

The Doppler shift for superluminal speeds is

$$\nu = \nu_0(\beta^2 - 1)^{½}/(\beta \cos \theta' - 1) \tag{20.6}$$

This can be seen by considering a baryonic plane wave, which is a combination of

$$\cos[(k\cdot x - \nu t)/2\pi] \qquad \text{and} \qquad \sin[(k\cdot x - \nu t)/2\pi] \tag{20.7}$$

Upon transforming from the uniship coordinate system, for example, to a coordinate system moving in the "x-direction" at a speed faster than light, both the energy ν' (up to a constant) and the time t' obtain a factor of i (that cancel each other) so eq. 20.6 is the correct frequency in the superluminal (faster than light) frame. The sign of the frequency is always positive by convention due to the form of baryonic waves and eq. 20.4 dictates the form of the denominator in eq. 20.6.

For large β >> 1 eq. 20.6 becomes approximately

$$\nu \approx \nu_0/\cos \theta' \tag{20.8}$$

In the forward direction θ' = 0 the Doppler shift goes to zero. Due to eq. 20.4 the maximum value of the Doppler shift for large β in the field of vision is

$$\nu \approx \beta\nu_0 \tag{20.9}$$

So the "wide" angle baryonic waves are shifted to large frequency.

Eq.20.6, and the discussion that follows, suggest that frequency shifts will be substantial for extremely fast uniships. The result will be a distorted view of the Megaverse.

However electronic imaging techniques can again be implemented to restore the "correct" view of the baryonic radiation. The combined effects of aberration and the Doppler shift on the view of the Megaverse from the uniship can be electronically corrected to give a "normal" view of the Megaverse. In addition a projection system, probably based on holograms, is required to transform (D – 1)-dimensional views of the Megaverse into sets of 3-dimensional depictions of parts of the (D – 1)-dimensional view.

20.3 Uniship Navigation

Navigating on earth and in space is often a difficult task. First one must know where one is and then one must know where the destination is, and how to get there. In the Megaverse all three items are challenging to discern. For we are in a D-dimensional space where our intuition, based as it is on three dimensional space, fails. We are thus at the mercy of technology to detect these three things with only electronic baryonic eyes to see and guide our motion.

Baryon detectors on board detect other universes through their baryonic radiation just as we detect stars and galaxies by their electromagnetic radiation currently. Within the next 50,000 years we anticipate that baryonic radiation optics will develop and mature to the point where it can provide visual capabilities similar to electromagnetic light that we use at present. Then we can develop universe maps for the Megaverse just as we have star maps currently.

An important issue is the ability to distinguish anti-matter universes from universes dominated by matter such as our universe. We do not want uniships to enter anti-universes unless we can properly shield them from disintegrating under particle-antiparticle interactions.

The navigation system, using 3-dimensional views (holograms) obtained from (D – 1)-dimensional pictures of the Megaverse, can then plot courses to universes of interest for exploration.

The courses selected then direct the (D – 1)-directional thrust system to execute the correct combination of accelerations to travel to the selected universe.

Most of the technology that we have discussed remains to be created. However the rapid progress of technology, if it continues, would seem to be able to provide the needed components in the future.

20.4 Uniship Exit from a Universe

We have seen that only baryonic radiation can penetrate a universe horizon. We must now amend that conclusion to consider the exit or entry of uniships from a universe. In the slingshot mechanism for exiting, and the other possible exit methods that we considered, a key role was played by the baryonic interaction in all cases. The role is easily visualized by considering the example of escaping from 2-dimensional flatland using magnetic force.

However in the current situation (4-dimensional object exiting into a D-dimensional space) we have to inquire into the nature and results of the exit. It is clear that the escape of a uniship must be view as a bubble (the 4-dimensional universe) subdividing into two 4-dimensional bubbles in (D – 1)-dimensional space in a continuous process rather like biological cell division. One bubble is of course the universe. The other bubble is the uniship which is also a surface from the viewpoint of the Megaverse. Fig. 20.3 symbolically depicts an intermediate stage in the subdivision process.

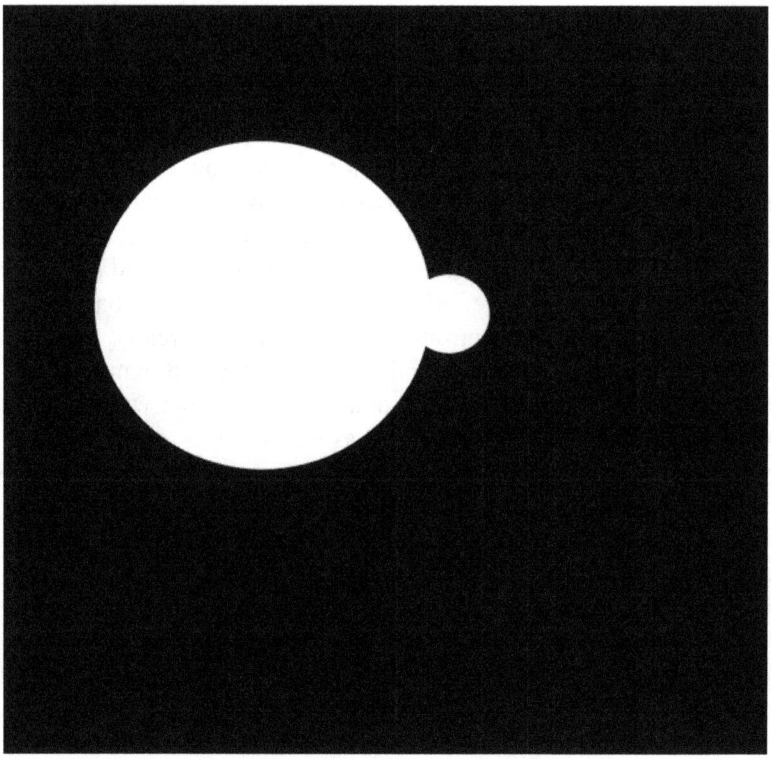

Figure 20.3. Symbolic depiction of an intermediate stage in the subdivision representing the exit of a uniship from a universe. The uniship (small circle) emerges from the universe (large circle) into the Megaverse. (Not drawn to scale.)

We thus see that a uniship is a 4-dimensional entity but with D - 1 thrust ports extruding into (D – 1)-dimensional space. (The thrust ports were generated by tidal effects in the slingshot process.) Thus it is a hybrid with 4-dimensional and D-dimensional parts. The crew, the equipment and the fuel modules will be in 3-dimensional space. The combustion chamber and thrust ports and viewing mechanism for navigation will be in (D – 1)-dimensional space. The combined parts will constitute the uniship. (We note that the 3-dimensional parts are in a subspace of (D – 1)-dimensional space just as a piece of paper is a 2-dimensional object in 3-dimensional space.) Therefore there are no connection problems between the 3-dimensional and (D – 1)-dimensional parts of the uniship.

20.5 Uniship Entry into a Universe

We now consider the entry of a uniship into a universe, which we will assume is 4-dimensional like our universe for reasons given earlier. The uniship has both 3-dimensional and (D – 1)-dimensional spatial parts. If we position the uniship 3-dimensional part within the universe and retract or rotate the (D – 1)-dimensional part into the 3-dimensions of the universe, then the uniship will be entirely within the universe giving a successful entry.

Retracting (D – 1)-dimensional parts is easy to visualize but requires a mechanism for retraction in (D – 4) of the (D – 1) dimensions. This mechanism can be a simple retraction mechanism but must exist in each of the dimensions outside the universe being entered. Rotating thrust exhausts also is easy to visualize but, again, the rotation device must be in the (D – 4) external dimensions and rotate into the three spatial dimensions of the universe being entered. In both cases we are faced with Cheshire cat situations: after the retraction or rotation the (D – 4)-dimensional devices that perform these chores will still exist and thus not be fully within the universe.

Whether the uniship can explore a new universe, being partly outside it, is an open question. This author feels that the uniship will be able to successfully navigate in the new universe rather like a shark swims the sea with its fin sticking out of the water. The (D – 4)-dimensional “fin” will add to the mass of the uniship but will not be affected by gravitation because it is out of the universe.

/

21. Issues for Life on a Megaverse Uniship

Life on a uniship that travels to universes is similar to life on a starship that travels to stars and galaxies except it has significantly more stringent requirements. Travel within the universe is measured in light years ranging to millions of light years. Travel within the Megaverse is most likely in hundreds of billions to trillions of light years in starship time. This results in major large requirements for fuel, materials strength and longevity, and in human suspended animation time, among other requirements.

Starship requirements are described in Blaha (2013a), 2014b) and (2014c). In this chapter we will describe some uniship requirements. The following chapters describe other uniship requirements.

21.1 Long Distance Starship Requirements for Travel to Far Stars and Galaxies

If we wish to travel to long distances – up to trillions of light years eventually, then critical advances are necessary that could take up to 50,000 years.

We see the uniship effort as a long term exploration and colonization program in an ever widening ring around our universe. In this chapter we discuss many of the advances that would be needed.

A major problem of uniships is the rapid progress of time on a much faster than light uniship. If a uniship has a speed that is much faster than the speed of light, then the progress of time in the uniship is much faster than the progress of time on earth.[189] For example if the uniship is traveling 5,000,000 times the speed of light, then the increase in time on the uniship is 5,000,000 times the increase in time on earth. In an interval of one year of earth time, 5,000,000 years will have passed on the uniship.

The extraordinarily fast passage of time on a very fast uniship requires materials, equipment and engines to continue to work effectively for long periods of uniship time which is just as real on a uniship as earth time is real on earth. One cannot avoid the fact that the distance traveled by a uniship measured in light years is equal to the time of flight to cover that distance measured in years.[190]

---

[189] See p. 15 of Blaha (2011c): *All The Universe*.

[190] Neglecting the time required to accelerate the starship at the beginning and the time required to decelerate back to a "normal" speed of a few miles per second.

Thus we come to the first important long distance uniship requirement – very long lifetime equipment and uniship superstructure. Other requirements follow in this chapter.

21.2 Long-Lived Materials

The materials that we use today to build large vehicles such as oil tankers, submarines and aircraft carriers are meant to last up to, at most, a century and often much less. Many of these materials age, deteriorate, rust, migrate within computer chips over time, and actually slowly flow like a liquid in many cases.

Not many materials keep their original characteristics over long periods of time. In the past fifty years there has been much progress in developing new harder, stronger and age resistant materials and metals. But uniships requirements are extraordinarily larger.

Uniships, in which time moves quickly so that thousands and perhaps millions of years of uniship time elapse, must be composed of materials with a very long stable lifetime. An important part of the R&D for a uniship is the development and use of age tolerant materials. The examination of materials used hundreds of thousands of years ago such as tools and dwellings shows the ravages of age. A uniship should have an initial goal of tens of millions of years of stability without aging. Ultimately one would hope that uniships that don't age in trillions of years could be built to travel to other universes.

These design requirements are far ahead of current technology.

21.3 Long-Lived Machinery and Electronics

If one has materials that preserve their composition, shape and performance characteristics over millions of years or more, then one can construct machinery and electronic gear such as computers that can last similar periods of time. Long-lived machinery and electronic gear then can be used when a uniship travels the Megaverse.

Long distance uniships need materials and machinery that last "nearly forever" – exactly the opposite of the intent of Earth industries.

21.4 Long Shelf Life Nuclear Reactors and Nuclear Shuttles

Several types of nuclear reactors are required for a uniship:

1. A continuous running reactor that can run for up to hundreds of millions of years to provide power to a uniship in flight to a distant location. This reactor may be a low power reactor. It should have a very long lifetime. That this is possible is suggested by the natural nuclear reactor that ran for millions of years about a billion years ago in the Congo.[191]

[191] The author suspects that the vast diversity of life in Africa may in part be due to genetic changes caused by this natural reactor. The development of mammalian life may in part also be due to this reactor and the radiation from its waste products and radioactive deposits in the Congo region over the millennia.

2. Long shelf life reactors that are not activated until a uniship destination is reached. These would power the uniship inside universes and their solar systems, and nuclear shuttles for travel and landings within a solar system.

21.5 Suspended Animation for Long Trips

It is necessary to have suspended animation available for crews on uniship journeys. With suspended animation a crew could go on a journey lasting hundreds of millions of years, or more, of uniship time, and, upon return to earth, have aged physiologically only a short time while out of suspended animation exploring distant universes and their star systems. The round trip travel time will not have aged them. When they return to earth they may be some months older, but their families and friends (having aged only by the earth travel time) will still be roughly contemporary with them.

A mechanism for long term suspended animation is thus a major requirement. Any suspended animation mechanism must take account of three important facts: 1) suspended animation must reduce human body temperatures to a low value to "halt" life processes and bodily decay; 2) lowering body temperatures will cause cells to rupture due to the expansion of water upon freezing; 3) the entry into suspended animation and the reentry to a normal bodily state must be rapid and uniform throughout the body.[192]

A mechanism to achieve these goals is not presently known. The current approaches to suspended animation (which all include lowering body temperature) are:

1. Replacing part or all of the blood in an organism with an "antifreeze" solution that will prevent cells and body tissue from bursting when the temperature is lowered. Revival takes place by raising the temperature of the organism while returning blood to the organism's circulatory system. This approach has been successfully applied to dogs that have been put into suspended animation for three hours. Unfortunately some of the dogs had nerve and coordination problems after revival.[193]

2. An organism can have a chemical injected or absorb a chemical while breathing that will counteract the tendency of water to expand when body temperature is lowered and/or lower the metabolic rate of the organism.

3. NASA and other groups have studied the possibility of placing humans into hibernation. Since hibernating organisms do age – perhaps more slowly – this approach is not true suspended animation.

4. A combination of electromagnetic "vibration" of a body having an innocuous chemical dispersed in the body (while awake or in suspended animation) might allow bodily

[192] One cannot "unfreeze" part of a human body and have the rest still frozen.
[193] At the University of Pittsburgh's Safar Center for Resuscitation Research.

temperatures to be lowered to a stable "frozen" state without cell rupturing. Turning off the electromagnetic vibration combined with a revival jolt might be an effective way to exit suspended animation procedure.

21.6 Robotic Driven Uniships

The initial uniship flights could be manned by robots rather than humans. This approach would be useful to test uniships, and their components, without endangering a crew. The robot guidance systems would, of course, have to be constructed of long-lived components. If it is successful then a robotic trip would also help demonstrate the long term reliability of long-lived computer equipment.

Robotic flights would be especially useful if a method of rapid faster than light, or instantaneous, communication between the uniship and earth existed. (See chapter 19.)

21.7 Long-Life Computer Chips

Computers hardened for battle and bad weather conditions currently exist. A long distance uniship would require computers with working lifetimes of between thousands and hundreds of millions of years. In time periods of these lengths computer chips would be subject to aging processes such as the intermixing of the metals composing the various chips of the computer and the aging of the wiring of the computer. Since new materials of greater strength and other superior properties are being discovered fairly frequently one can hope that the required types of metals and materials will eventually be found.

21.8 Space Dust

The effect of dust and gas molecules in space on uniships are of great importance. These effect should be detectable in "short" distance uniship voyages. If it is important, as it seems to be, then the design of shielding for long distance uniships should incorporate appropriate "armor" to protect the uniship and crew.

21.9 Length Dilation Effect

Lengths on a uniship, traveling at high speed much greater than the speed of light, are significantly dilated. A length measured on a uniship will appear to be larger to an observer on earth by a factor of the speed measured in terms of the speed of light than the length on earth. For example if a uniship is moving at 5,000 times the speed of light then a 2 meter long stick on the uniship would appear to be 10,000 meters long to an earth observer.[194]

Does this length dilation phenomenon affect the contents of the speeding uniship? No. It is an illusion that the earth observer "sees." An occupant of the uniship would not notice a

[194] This discussion assumes that the stick and the starship motion are parallel. For a detailed discussion of this length contraction phenomena see Blaha (2011c) p. 17.

difference and would see the stick as still two meters in length. Due to length dilation, a starship traveling at speeds much beyond the speed of light, would appear to be enormously long.

21.10 QFT Acceleration Thermal Vacuum Heating – Particle Baths

Recently questions have been raised about spaceships accelerating at a high rate in 'empty' space. Based on an analysis of Unruh and others[195] it has been suggested that the ship would see an incoming wave of particles and a 'heated' vacuum due to its motion. The basis of this claim and the work of Unruh (and others) is standard quantum field theory, which in turn is based on Special Relativity.

Since accelerating reference frame transformations and other exotic reference frame transformations are outside the framework of Special Relativity we suggest that this spaceship phenomena merely reflects a flaw in the choice of second quantization. In section 7.6.3 and chapter 8 (as well as Appendix A) we describe a more general formulation of quantum field theory in which particle states are unchanged under transformations between accelerating reference systems as well as under transformations to exotic coordinate systems. Thus the particle 'baths' suggested by conventional quantum field theory do not take place for accelerating spaceships nor does the vacuum 'heat up' and spew particles.

[195] See the relevant references in Appendix A.

/

22. 'Early' Space Transportation

The vehicles for voyages to space have primarily been via rockets. Recently a new generation of privately built rockets have been developed which promise to provide a more cost effective in transporting cargo into space. In addition the US Navy has developed and is planning to deploy electromagnetic and rail guns for combat. These types of guns and the earlier big guns like Big Bertha (a World War I German gun) are capable of being engineered into space guns to cheaply loft cargo into space. Blaha (2013a) discusses the possibilities of non-rocket cargo transport to space – *particularly its relative inexpensiveness*. We summarized some of the possibilities for space technology discussed in Blaha (2013a), which is presented below.

22.1 Rocketry and Space Guns for Space Flight in Blaha (2013a)

The possibilities and issues for new vehicles for solar system travel can be partially summarized by the list:

MULTI-STAGE SPACE GUNS FOR COST-EFFECTIVE CARGO SHIPMENT TO NEAR SPACE

Single-Stage Space Guns
Multi-Stage Space Guns
Basic Multi-Stage Space Gun
Enhanced Multi-Stage Space Gun
Other Measures To Minimize Space Gun Costs
Moving Massive Amounts Of Cargo Into Space

NUCLEAR ROCKETS: THE PATH TO THE PLANETS

History Of Nuclear Rocket R&D
Bimodal Nuclear Thermal Rockets
Nerva Program Overview
Proposal For Nuclear Rocket Mars Missions Vs. Ultra-Large Chemical Rockets
New Russian Nuclear Rocket Program
Pulsed Pellet Micro-Nuclear Explosion Rocket Drives
Pulsed Pellet Micro-Nuclear Explosion
Pellet Acceleration Methods
Suggested Initial Flights Of A Nuclear Rocket

/

23. Uniship Development Time Frame

The development and deployment of uniships for the exploration of the Cosmos will be a supreme technological accomplishment of the human race. It will lead to an extraordinary growth of human culture to embody a combination of the best attributes of humanity enriched by contact with the cultures of the many extraterrestrial civilizations[196] scattered across the Megaverse.

23.1 The Development Phases

Although it is difficult to forecast how a complex development project will take shape, especially when so many parts of it require new technology in many areas, we shall make a tentative schedule of phases knowing that chance and difficulties will probably cause the actual schedule to differ. In making this schedule we will assume that major unforeseen breakthroughs will not occur. If a major breakthrough occurs such as "warp drive" then the development schedule would change drastically. However the author believes breakthroughs of that sort will not happen. Rather it is more likely to be a long term, expensive, hard slog towards eventual success.

We expect the following major phases to occur:

1. Development of efficient, large scale cargo and people transport to earth orbit over the next twenty years.

2. Manned inner planet exploration in the following thirty years using nuclear and possibly fusion powered space ships.

3. Manned exploration of the outer planets and moons and possibly the development of colonies on Mars and outer planet moons for scientific and commercial purposes. The time for this phase is probably another forty years following (but overlapping) the second phase.

4. Concurrent development of faster than light starships over a period of 150 years together with suspended animation and other technologies. The author believes the only viable propulsion approach is a quark-gluon ion drive described in Blaha (2013a) and earlier books. This approach would enable starships to evade the speed of light limit.

---

[196] Blaha (2011b), (2013a).

5. Assuming success in phase 4 an exploration and colonization phase of stars within a hundred light years of earth would follow with a duration of 1,000 years.

6. Technological advances in starships should then make exploration of the galaxy and nearby galaxies possible. This phase could last for many thousands of years and probably be extended to tens of thousands of years if economically and scientifically justified. Starships would continue to be propelled by quark-gluon drives using fusion energy.

7. Concurrent with phase 6 uniship design and development should take place using energy from particle-antiparticle annihilation – the most efficient energy source known. The many parts of this development will probably require many tens of thousands of years to put together. We suggest a ballpark figure of 50,000 years. A problem of major importance is to develop a method to reach extraordinary speeds of the order of millions of times the speed of light with uniship occupants experiencing humanly bearable accelerations.

8. After the design, development and testing of a uniship within our universe and in the nearby Megaverse an exploration program for the Megaverse can commence with the exploration of "nearby" universes. This exploration phase will undoubtedly take tens of thousands of years up to millions of years depending on the benefits derived from exploration, and from perhaps meeting other civilizations.

23.2 Spaceship, Starship and Uniship Propulsion Phases

The different "ages" of space travel are best characterized by propulsion mechanisms and energy sources just as transportation on earth can be characterized by animal power (horses, etc.), steam powered engines, oil-powered engines, and jet engines.

The ages of space travel can be characterized as

1. The age of chemically powered spaceships

2. The age of Nuclear/Fusion Powered Spaceships (under development)

3. The age of Quark-Gluon powered Starships (at an initial experimental stage at CERN in particle physics high energy ion-ion collisions)

4. The age of particle-antiparticle powered Uniships (a gleam in the eye of Megaverse enthusiasts)

23.3 Cost Issues

The starship development program mentioned above is a necessary precursor to the uniship development effort. The basic initial framework for the uniship project is the starship development. Uniships have similar design needs for the most part but on a much larger scale. Uniships will need nuclear and/or fusion engines for maneuvering in the solar systems of other universes. They may also need quark-gluon ion drive engines for low speed (but much greater than light speed) travel within a universe. Lastly they will need the most powerful energy source that we know of: particle-antiparticle powered engines for travel across the vast distances between universes.

In Blaha (2013a) we estimated the cost of design and construction of a faster than light starship at a trillion current dollars spread over one hundred years. In view of the continuing financial problems of the United States and Europe it is likely that the development time will be longer – perhaps one hundred and fifty years.

The development and construction of a uniship can build on the knowledge gained in the starship program. However its much greater requirements, and size, suggest that its cost will be in the five to ten trillion dollar range in current dollars.

Undoubtedly the research and development efforts in these programs will have tremendous technical spinoff benefits which will enrich world technology just as the space programs have done in the past. And the successful exploration of first the stars and galaxies of our universe, and then other universes should yield an enormously valuable return on our investment.

Appendix A of Blaha (2014a) describes our space and starship design and dynamics for the expansion of Man into the solar system and the galaxies of our universe. Appendix B describes the uniship design, and major considerations, for trips into the Megaverse. So we look to the future with confidence in the eventual expansion of humanity into the Megaverse.

23.4 Constant Superluminal Starship Travel

Assuming a starship has accelerated to an enormous *real* speed such as a speed between 5000c and 30,000c we can turn off the superluminal engines. The starship then moves at this constant speed in the absence of other forces, gravity, retarding effects of space dust, and so on.)

Consider a starship speed of 5000c. Any place in the galaxy is a short travel time away. And nearby galaxies are reachable as well. Fig. 23.1 shows the time required to reach various interesting destinations at a much higher speed of 30,000c.

Destination	Distance (ly)	Approximate Travel Time (years)
To the other end of the Milky Way Galaxy	100,000	3
To the Center of the Milky Way	30,000	1
Large Magellenic Galaxy	150,000	5
Small Magellenic Galaxy	200,000	7
Andromeda Galaxy	2,000,000	70

Figure 23.1. "Coasting" part of travel time to various destinations at a real velocity of 30,000c.

Since much, much higher "coasting" velocities are also possible almost the entire visible universe becomes accessible to Mankind if we can boost quark-gluon exhaust velocities to very large values. Mankind then has an incredible future if it has the will to seize it.

23.5 Deceleration of a Tachyonic Starship to Sublight Speeds

Eventually all journeys end, so we will now examine the deceleration of a starship as it approaches its destination. We turn on the superluminal engine. The thrust is reversed ($g \rightarrow -g$) to decelerate as the target star system is approached.

23.6 Fuel Consumption

The acceleration of a rocket of mass m with a propellant exhaust speed v_e in the rocket's rest frame is given by

$$dv'/dt' = (v_e/m)\, dm/dt' \qquad (23.1)$$

and thus the constant g of eq. A.1 is

$$g = mdv'/dt' = v_e\, dm/dt' \qquad (23.2)$$

Since we intend to generate the thrust with a quark-gluon plasma producing an extremely high-energy exhaust we will *choose* the value of the starship acceleration to be equal to the acceleration due to gravity at the earth's surface g_E times $8(1 + i)$:[197]

$$g/m = 8(1 + i)g_E = 8(1 + i)980 \text{ cm/sec}^2 \qquad (23.3)$$

where m is the mass of the starship. If we specify an exhaust velocity v_e

[197] Eight g's in astronaut terminology.

$$v_e = -1000(c + ic) \quad (23.4)$$

which is a reasonable choice for the exit speed thrust of the fireball then

$$dm/dt' = -2.61\times10^{-10}\ m \quad (23.5)$$

If the starship weighs 10,000 metric tons[198] then

$$dm/dt' = -2.61\ gm/sec \quad (23.6)$$

From the viewpoint of rockets, dm/dt' is a small quantity. But, due to time dilation, the cumulative effect of dm/dt' in multi-year travel in starship time is a relatively large amount of fuel.

However if a starship can use processed material from asteroids and moons to make fuel then the limitation on travel imposed by fuel consumption can be circumvented. The fuel need not be composed of specific elements such as lead or uranium but could be spherules composed of a variety of materials if the starship engine were designed to handle such a variety of spherules.

The amount of fuel used per unit time would appear to be acceptable for quark-gluon plasma production for an ion drive. Currently minuscule amounts of plasma are created with ion-ion collisions. Colliding spherules of 1 milligram mass would require a not unreasonable collision rate of 1305 nominal collisions per second.

We have seen that a starship with tremendous capabilities for exploring the universe can be built if we can build a quark-gluon ion drive that produces large complex accelerations.

198 About one-fifth the mass of the ship Queen Elizabeth.

24. Uniship Particle Annihilation Drive

This chapter (from Blaha (2011a)) epitomizes the possibility that the CERN LHC can be used as a *test instrument for the development* of a starship ion drive capable of travel to the stars (after the LHC is retired from its investigation of elementary particle phenomena).

We have shown in I that quarks and gluons have complex-valued velocities that enable them to travel faster-than-light (tachyonic), which we showed in I enables quarks to generate more massive particles after collisions than the initial particles. Recently, new experiments (announced in April, 2017[199]) at the LHC have shown that ultra-high energy proton-proton collisions have producd an abundance of particles similar to that produced in collisions of atomic nuclei. Thus initial quarks in protons effectively 'multiply' to have atomic nuclei characteristics – a feature we attribute to their tachyonic nature.

The tachyonic attributes that seem to explain these new results suggest that a faster-than-light ion starship engine could be created. CERN LHC seems able to do design studies to aid in the development of this new type of starship engine at now accessible energies.

---

[199] ALICE Collaboration, reported in the journal Nature Physics (April, 2017).

25. Voyages into the Megaverse

Traveling in in the Megaverse between universes places extraordinary demands on uniships. Speed, acceleration, and fuel requirements are the primary issues because the distance between universes is vast and human lifetimes are short. We should like to be able to travel fairly quickly between universes. If one wishes colonization, commerce, and timely exploration then the trip between two universes, on average, should be perhaps six months to a year of earth time. (Time on a uniship proceeds much, much more rapidly than earth time. But we can circumvent this potential problem using suspended animation for passengers and very long-live machinery for the uniship and its contents.)

In this chapter we will consider Uniship distance and speed related requirements for "short" distance travel in the Megaverse.

25.1 The Distance Scales of Our Universe and the Megaverse

We will begin with a consideration of distance scales between galaxies in our universe and anticipated distance scales between universes in the Megaverse. Our universe has a web of groups of galaxies. Typically the distances between galaxies in a group are of the order of several million light years.[200]

The distances between universes in the Megaverse are a matter of conjecture. However if we take the order of magnitude of the ratio of the size of a galaxy (say the Milky Way which is 100,000 light years in diameter) to the separation of galaxies in a group of galaxies (say three mllion light years) as a guide (a factor of 30) and use the same ratio with the size of our universe as the input, then, with the size of the (visible) universe being about 50 billion light years the order of magnitude of the relative separation between universes could be roughly estimated to be perhaps two trillion light years.

This estimate is at best an order of magnitude estimate. It suggests that Megaverse universe distances are of the order of 1,000 times distance scales in our universe – a not unreasonable value.

25.2 Starship Distance and Speed Requirements in Our Universe

In our book *All the Universe!* (and in earlier books) we developed the theory of faster-than-light starships for travel between stars and galaxies in our universe. For the reader's convenience we reproduce part of *All the Universe!* It appears that speeds of 60,000c give

[200] BOSS – Baryon Oscillation Spectroscopic Survey and other studies of WMAP data.

acceptable travel times (up to a year) within our galaxy, and *speeds of a few million c give acceptable travel times to nearby universes where c denotes the speed of light.*

At 3,000,000,000c a universe that is 2,000,000,000 light years away could be reached in eight months (neglecting acceleration and deceleration times) – an acceptable time for trade, exploration and possibly colonization. The fundamental problem is to develop an energy source that can fuel such enormous speeds – a reason for anticipating a 50,000 years of necessary development time.

Destination	Distance (ly)	Approximate Travel Time (years)
To the other end of the Milky Way Galaxy	100,000	3
To the Center of the Milky Way	30,000	1
Large Magellenic Galaxy	150,000	5
Small Magellenic Galaxy	200,000	7
Andromeda Galaxy	2,000,000	70

Figure 25.1. "Coasting" part of travel time to various destinations at a real velocity of 30,000c.

25.3 Uniship Distance and Speed Requirements for "Short" Distances in the Megaverse

In this section we will consider some issues associated with uniship "short distance" trips in the Megaverse such as a three trillion light year trip to a nearby universe.[201] The primary issue is the energy (and fuel) required to make a trip at high speed so that the travel time is of the order of months. Another important issue is the acceleration times that are required to attain high speeds.

25.3.1 Energy Required to Attain High Velocities

The energy E required to reach a high velocity much greater than the speed of light is not a simple mathematical expression in the speed v because the energy of an object moving much faster than the speed of light approaches zero. (The momentum approaches mc where m is the mass of the object. Thus $E^2 - c^2p^2 = -m^2c^4$ for $v > c$. The object is tachyonic.)

We will calculate the energy required to reach a speed $v > c$ in the earth's rest frame using the development presented earlier in this book. The energy E expended in the spatial x interval from x_0 to x is defined as

[201] We will use the results found earlier for acceleration and velocity in the x direction. The general case follows directly from this special case.

$$E = \int_{x_0}^{X} F dx = \int_{t_0}^{t} F v dt \quad (25.1)$$

where t_0 is the initial time and t is the final time. From 25.1 we obtain

$$E = \int_{t_0}^{t} F v dt = \int_{t_0}^{t} g \gamma v dt \quad (25.2)$$

Eq. 25.2 can be transformed to the form:

$$E = (m/2) \int_{v_0}^{v} dv \, \gamma^{-1} \, d[(\gamma v)^2]/dv \quad (25.3)$$

The exact form of the integral is:

$$E = [m\gamma(v^2 + c^2)/2 + mc^2/(2\gamma)] - [m\gamma_0(v_0^2 + c^2)/2 + mc^2/(2\gamma_0)] \quad (25.4)$$

where $\gamma_0 = (1 - v_0^2/c^2)^{-1/2}$. Since the velocity is in general complex-valued, v and γ are also complex-valued as is E.

Using the Reality group that we introduced in previous books, the physical velocity is the absolute value of v, |v|, and the physical value of E is |E|. |E| is calculated by first substituting the complex values of v, γ, v_0 and γ_0 in eq. 25.2 and then taking the absolute value of the resulting complex quantity E.

We begin by defining

$$E_{part}(v) = m\gamma(v^2 + c^2)/2 + mc^2/(2\gamma) \quad (25.5)$$

The integral in eq. 25.2, although superficially real-valued has complex quantities in the integrand. The result of the integration from a speed below the speed of light to a speed greater than the speed of light can be written as

$$E_{tot} = E_{part}(c - i\varepsilon) - E_{part}(v_0) + E_{part}(v) - E_{part}(c + i\varepsilon) \quad (25.6)$$

as $\varepsilon \rightarrow 0$ due to the singularity at v = c. Since we are interested in the energy required to reach ultra-high speeds of the order of tens of thousands to trillions of times the speed of light, we see that

$$E_{tot} \rightarrow \lim_{v\rightarrow\infty} E_{part}(v) = -imcv/2 \quad (25.7)$$

to leading order in v. Thus

$$|E_{tot}| \rightarrow mc|v|/2 \quad (25.8)$$

or, expressed in a more convenient way:

$$|E_{tot}| \rightarrow \frac{1}{2} E_{rest} |v|/c \tag{25.9}$$

as $v \rightarrow \infty$ in the earth's reference frame where $E_{rest} = mc^2$ is the rest mass-energy of the entire ship including the fuel. Knowing the desired cruising speed v of a starship or uniship the ship must have an energy supply equal to $2|E_{tot}| = mc|v|$ when it leaves the earth's reference frame to provide for acceleration *and deceleration* plus the additional energy needed for other ship functions. For a round trip an energy supply of $4|E_{tot}| = 2mc|v|$ would be needed for the ship's engines.

25.3.2 Acceleration Time Required to Reach Extremely High Speed

We can determine the acceleration time required to reach velocity v from eq. (25.9 of the previous section:

$$v = c\{1 - 2/(1 + ((c + v_0)/(c - v_0))\exp[2g(t - t_0)/(mc)])\} \tag{25.9a}$$

Inverting eq. 25.9 above yields the acceleration time interval required to achieve a speed v in the earth's reference frame:

$$t(v) - t_0 = (mc/2g)\ln\{[(c - v_0)/(c + v_0)]\ [(c + v)/(c - v)]\} \tag{25.10}$$

where v_0 is the speed at t_0. For large v in the earth reference frame the time interval approaches

$$|t(v) - t_0| \rightarrow (mc/2g)\ln[(c - v_0)/(c + v_0)] + mc^2/(gv) \tag{25.11}$$

to leading order in v. As $v \rightarrow \infty$ the interval becomes a constant:

$$|t(v) - t_0| \rightarrow (mc/2g)\ln[(c - v_0)/(c + v_0)] \tag{25.12}$$

The reason for this limit on the earth reference frame time interval can be understood when one realizes that the corresponding ship time interval approaches infinity. Ship time increases much faster at high speeds greater than $\sqrt{2}c$. We shall see this in the next subsection.

25.3.3 Speed, Distance and Acceleration Time in the Starship/Uniship Reference Frame

In this subsection we will calculate dynamical quantities from the point of view of the starship/uniship reference frame.

In case examined in the previous section we found the acceleration in the earth reference frame to be $\gamma g/m$ by eq. 25.3. In the ship reference frame the acceleration is g/m. See the quoted section 25.6 within the previous section for more detail.

The transformation law from the earth reference frame (unprimed coordinates) to the starship/uniship reference frame (primed coordinates) in which the ship is moving at instantaneous speed v in the x direction is

$$\begin{aligned} t' &= \gamma(t - \beta x/c) \\ x' &= \gamma(x - \beta ct) \\ y' &= y \\ z' &= z \end{aligned} \tag{25.13}$$

Substituting

$$x = x_0 + (mc^2/g)\ln[(1 - v_0/c)/(1 - v/c)] - c(t - t_0) \tag{25.11a}$$

from the previous section and

$$t(v) - t_0 = (mc/2g)\ln\{[(c - v_0)/(c + v_0)]\,[(c + v)/(c - v)]\} \tag{25.10}$$

from this section with $t_0 = 0$ and $x_0 = 0$, we obtain the ship time to be

$$t' = \gamma\{(m(c - v)/2g)\ln\{[(c - v_0)/(c + v_0)]\,[(c + v)/(c - v)]\} - (mv/g)\ln[(c - v_0)/(c - v)]\} \tag{25.14}$$

Eq. 25.14 gives starship time as a function of speed. Thus the faster a ship speeds the more quickly time passes on the ship. As a result ships should have long lifetime equipment and place passengers in suspended animation for long periods of time.

For large v, the absolute value of the ship time approaches

$$|t'| \rightarrow |mc/g)\ln v| \tag{25.15}$$

Thus the ship time approaches infinity as the ship speed approaches infinity. This situation explains the limit on earth time in eq. 25.12. It corresponds to ship time approaching infinity. (The ship speed approaches infinity in this limit as well.) Of course the limit is never reached because it would require an infinite amount of thrust and fuel. However it allows very, very large velocities to be reached making starship and uniship travel in reasonable time frames possible.

If the conditions in the above section hold, then it is possible to reach "infinite" speed in a finite time in the earth's reference frame. To reach this limiting speed will require an infinite amount of starship time, and fuel.

26. The Dimensions of the Parts of a Uniship in the Megaverse

A uniship in the Megaverse has parts with different dimensionality. Some parts are 3-dimensional; some parts are (D – 1)-dimensional. We will examine the dimensionality of the various uniship parts in this chapter. Fig. 26.1 displays the parts of a uniship crudely mapped to two dimensions.

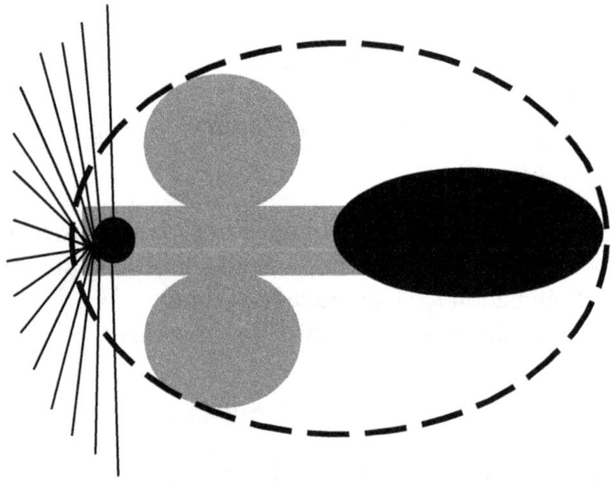

Figure 26.1. A 2-dimensional projection of a uniship with the actual dimensions of each part in the Megaverse marked. The dashed ellipse encloses the 3-dimensional part of the uniship. The white space within the ellipse is (D – 1)-dimensional Megaverse space. The thrust tubes (spokes) are partly in the 3-dimensional uniship bubble. The outer parts of the thrust tubes are in (D – 1)-dimensional Megaverse space pointing in the various Megaverse directions.

26.1 Dimension of a Uniship in the Megaverse

From an overall perspective, a uniship is a 3-dimensional irregularly-shaped bubble in the (D – 1)-dimensional Megaverse space. We take the time dimension to be the same in our 4-dimensional universe (and thus the uniship) and in the D-dimensional Megaverse space-time.

This is the simplest choice of times. More complex choices of time in each case are possible. But the physics in any case will remain the same.

Since the space-time in the uniship is the same as our space-time the physical processes will be the same as they would be in our universe.

The thrust tubes are somewhat different. They are presumably long – of the order of kilometers so that the slingshot into the Megaverse will swing them in the D - 1 spatial directions of the Megaverse. However, being attached to the 3-dimensional uniship they will be 3-dimensional at their beginning at the uniship end, and transition to (D – 1)-dimensional beyond that point. Despite this change of dimension they will each still be linear –not curved. A simple geometric example of this is a line attached to a circle's edge – all in the x-y plane. If the end of the line is swung into the z direction by some means, then one end will be in the x-y plane and the other end pointing in the z direction. The line remains straight.

26.2 Viewports of a Uniship into the Megaverse

Seeing is easy in our universe. Our eyes can turn in any direction and they are naturally adapted to seeing a range of electromagnetic radiation. In the Megaverse seeing is more problematic due to the many more dimensions, and the lack of electromagnetic radiation from universes.

A solution to these problems is:

1. To "see" using baryonic or electromagnetic radiation using devices that manipulate and transform radiation into visible radiation displayed on monitors.

2. To sense baryonic and electromagnetic radiation with detectors positioned on the end of thrust tubes. Each detector will detect radiation from all directions through a controllable rotating (D – 1)-dimensional mechanism. The radiation will be transformed into electromagnetic signals transmitted down a cable to the uniship body. There they will be combined like insects combine light from their eye facets into composite images that can be viewed and used for navigation by the crew. Thus the thrust tubes have a dual function: to drive the uniship and to gather radiation so that the crew can view the universes of the Megaverse.

Thus it is feasible to navigate in the Megaverse using radiation. There are significant technical hurdles such as the creation of small baryonic radiation detection devices, and the creation of the equivalent of mechanisms that manipulate and magnify electromagnetic radiation.

27. Life Forms in the Megaverse

We have seen that there is evidence for the existence of the Megaverse . We have also seen that it is quaite likely that matter and energy exist in Megaverse space between universes. It is possible that Megaverse matter might have accreted due to gravity and other forces to form stars and planets. If so, then possibly life exists in the Megaverse outside of universes. This possibility raises many provocative questions. We will call Megaverse life *Megaversian.*

We know that life on earth exists in many niches that we would consider hostile: deserts, high altitudes, deep oceans, intense cold and darkness, and even kilometers below the surface of the earth. Life is prevalent on earth.

We have found other solar system bodies have environments that could support life. And we have found planets around distant stars that appear to be capable of supporting life.

Given these facts it sems reasonable to inquire as to the form and nature of life within the D-dimensional Megaverse space.

27.1 Why Life?

If life is prevalent within, and without, our universe then there must be a fundamental reason for Nature's tendency to produce life in just about any 'hospitable' environment. We believe the reason is that "*Life is maximally entropic.*" Entropy almost always increases with time. Sometimes quickly; sometimes more slowly. In any given situation where life exists we believe that entropy increases more quickly than it would if life were not present. This hypothesis, of course, needs comprehensive study. But we will assume it as a working hypothesis in our investigation.

27.2 A Necessary Condition for Life

As part of our fundamental hypothesis for the existence of life, we require that energetic processes must exist in an environment where life appears.[202] Living things need energy to develop and grow as well as for motion should they be mobile.

If one considers the simple fact that giant living objects do not pop into existence, but rather start from the small, we see another requirement for life is that its origin is local – life originates in the small. On earth it went through an evolutionary process starting from viruses to cells to multi-cell creatures – eventuslly becoming the small and large creatures of our experience.

---

[202] See Feinberg (1980) for a study of energy processes that could support life.

But one must remember the beginning of all life is in the small and based on local sources of energy that could fuel life in the small and enable it to grow and evolve over time. *We conclude the beginnings of life require a local source of energy.In the earliest stages of the development of life forms, the seed ('virus'?) must utilize an energy source in its immediate vicimity. Only at 'later' stages can life forage for energy over a distance.*

27.3 Types of Known Life

A great deal is known about life on earth. However there is still much to learn. We will pursue the question of Megaverse life based on our knowledge of aspects of earth life.

27.4 Fundamental Aspects of Megaverse Life

It would be nice if the physical laws, coupling constants, and particle masses of the Megaverse were the same as those of our 4-dimensional universe. And it may well be so, since our Continuous Creation Model based on an inflow from the Megaverse (Chapter 14) would suggest it.

We shall assume that this is the case but, where possible, not rely on precise details based on phenomena in our universe. Our focus will be on the impact of D dimensions on features of Megaverse life.

We will assume, therefore, that the stable nuclei of atoms are the same in our universe and the Megaverse. However the arrangement of electron shells, and their energies, around the nuclei of atoms will be different due to (D – 1)-dimensional space. As a result the chemistry of molecules and, most particularly, of the equivalent of DNA (if there is one) will differ substantially.

Thus we can expect life in Megaverse space to be quite different from life within our universe. Nevertheless, we can make some 'conclusive' statements based on general considerations.

27.5 Megaversian DNA

DNA is the common denominator of all known forms of life on earth. We assume an equivalent to DNA exists as the basis of life in (D – 1)-dimensional Megaverse space (Megaverse 'time' is the additional dimension to make the D-dimensional Megaverse.) Based on earth life, we assume Megaversian DNA, *Mega-DNA*, has the following properties:

1. It consists of long strands of linked nucleotides (the equivalents of cytosine, guanine, adenine, and thymine) as well as equivalents of deoxyribose and a phosphate group.

2. There are two strands, one of which may be viewed as a 'back up' copy, to provide stability and for use in DNA replication.

The purpose of the *two* strands suggests that the number of strands is independent of the dimensionality of space.

The large number of dimensions of Megaverse space has a number of important implications:

1. The long Mega-DNA strands can be compacted to better fit into cells as DNA strands are compacted in balls in cells on earth. Earth DNA strands range up to 2 meters in length. Mega-DNA strands, perhaps of length of the order of

$$2(\text{Megaverse-cell radius})^{(D-4)} \text{ meters}$$

   can be fit within Megaverse-life cells.

2. The additional Megaverse dimensions suggest more 'flexibility' in the Mega-DNA to provide more epigenetic adaptability and more capacity for mutations.

3. Mega-DNA should support a greater variety of cell and multi-cell life than found on earth.

Thus we can envision 'fantastic' creatures able to rapidly adapt to changing environmental conditions and to possibly have the feature called 'shape shifting' in popular SciFi movies.

27.6 Megaversian Brain Size

Perhaps the most important effect of a large number of dimensions is intelligence level. In our universe, assuming electromagnetic connections in the brain, the size of brains is constrained by the time it takes for signals to go between parts of the brain. (Currently many investigators believe that consciousness, and possibly abstract thought, are a result of total brain 'collaboration.') The maximum brain size is usually estimated to be slightly larger than that of the human brain.

In the Megaverse we do not have three space dimensions but rather D – 1 spatial dimensions. Thus the volume of a Megaverse brain is not proportional to r^3 where r is the radius of the brain. Rather it is proportional to $r^{(D-1)}$. Thus a Megaverse brain of the same radius as an earthly brain has incomparably more volume and contents than an earthly brain of the same radius. Thus if a Megaverse intelligent species should exist one can expect it would have massively more memory and analytical power.

Another aspect of Megaversian brain structure is the possibility of distributed brain power. We are familiar with networks of computers uniting to do massive computations. Some earthly species such as the octopus have a distributed brain structure: an octopus has a central brain and a 'sub-brain' in each of its eight limbs. Distributed brain power in Megaverse creatures could result in formidable computational and analytical abilities.

27.7 Megaversian Locomotion

In our universe, particularly on earth, larger animals tend to have two or four legs. Smaller creatures such as insects may have many more legs. Two-legged (and four-legged) animals appear to have two (or four) legs to be able to turn effectively in three dimensions. They do not have three or five legs because the added mobility is outweighed by the added stability requirements placed on the brain and nervous system.

In the Megaverse we would expect that creatures would have D – 1 legs (appendages) to maneuver in D – 1 spatial dimensions. It is possible that appendages would have 'sub-brains' like the octopus to off load processing and control of limbs. Whether they have hands and arms is open to question since legs could also play the role of arms.

27.8 Megaversian Vision

In D – 1 dimensions it would appear reasonable to have compound eyes like insect species on earth. Processing the data coming from the eyes would be a significant burden on the Megaverse brain.

27.9 Composition of Megaversian Life

If the elements present in Megaverse matter are similar to those in our universe the composition of creatures may be similar to that of life in our universe. However the many Megaverse dimensions, and the expected differences in the electronic structure of atoms and compounds might lead to a different chemical composition of life. Perhaps arsenic-based life might exist. Organic chemicals would be different in composition and structure.

These differences might lead to changes in the environments suitable for life: temperature extremes, atmospheres, and food requirements among other things.

27.10 Megaversian Societies and Civilizations

The average density of matter in the Megaverse may be expected to be low (section 14.14) and, consequently, the number of planets and stars per unit volume will be correspondingly small. On some of these objects life may develop and, on a much smaller number, societies and possibly civilizations may appear. After all, on earth we find complex societies of ants, bees, and so on. Their societies are similar to human socities in many respects[203] despite the vast difference between Mankind and insects. *The key factor in the growth of civilizations is the availability of large amounts of surplus energy.*

---

[203] See Blaha (2010c) for a comprehensive study of human societies and almost all the known civilizations of Mankind. It shows that the 'ups and downs' of civilizations and societies is based on energetics (Thermodynamics).

On this basis we suggest that Megaverse societies and civilizations are a likely possibility given the existence of Megaverse life forms.

27.11 Megaversian Life

If Megaverse bodies are at all habitable, then one can expect life to exist, and yet to be very different due to the D Megaverse dimensions. Yet despite the differences one can expect certain similarities to life in our universe.

Appendix A. The Local Definition of Asymptotic Particle States

This appendix reprints S. Blaha, "The Local Definition of Asymptotic Particle States", IL Nuovo Cimento **49A**, 35 (1979).[204] It describes the Pseudoquantization of boson and fermion field theories for use in the quantization of fields in universes and the Megaverse in chapter 7.

[204] © Copyright Stephen Blaha 1978.

Appendix B. New Framework for Gauge Field Theories

This appendix reprints S. Blaha, "New Framework for Gauge Field Theories", IL Nuovo Cimento **49A**, 113 (1979).[205] It describes the Pseudoquantization of gauge field theories for the purposes of defining higher derivative field theories and for use in the quantization of fields in universes and the Megaverse in chapter 7.

[205] © Copyright Stephen Blaha 1978.

REFERENCES

Akhiezer, N. I., Frink, A. H. (tr), 1962, *The Calculus of Variations* (Blaisdell Publishing, New York, 1962).

Bjorken, J. D., Drell, S. D., 1964, *Relativistic Quantum Mechanics* (McGraw-Hill, New York, 1965).

Bjorken, J. D., Drell, S. D., 1965, *Relativistic Quantum Fields* (McGraw-Hill, New York, 1965).

Blaha, S., 1998, *Cosmos and Consciousness* (Pingree-Hill Publishing, Auburn, NH, 1998).

_______, 2002, *A Finite Unified Quantum Field Theory of the Elementary Particle Standard Model and Quantum Gravity Based on New Quantum Dimensions™ & a New Paradigm in the Calculus of Variations* (Pingree-Hill Publishing, Auburn, NH, 2002).

_______, 2003, *A Finite Unified Quantum Field Theory of the Elementary Particle Standard Model and Quantum Gravity Based on New Quantum Dimensions™ and a New Paradigm in the Calculus of Variations* (Pingree-Hill Publishing, Auburn, NH, 2003).

_______, 2004, *Quantum Big Bang Cosmology: Complex Space-time General Relativity, Quantum Coordinates™ Dodecahedral Universe, Inflation, and New Spin 0, ½, 1 & 2 Tachyons & Imagyons* (Pingree-Hill Publishing, Auburn, NH, 2004).

______, *2005a, Quantum Theory of the Third Kind: A New Type of Divergence-free Quantum Field Theory Supporting a Unified Standard Model of Elementary Particles and Quantum Gravity based on a New Method in the Calculus of Variations* (Pingree-Hill Publishing, Auburn, NH, 2005).

______, 2005b, *The Metatheory of Physics Theories, and the Theory of Everything as a Quantum Computer Language* (Pingree-Hill Publishing, Auburn, NH, 2005).

______, 2005c, *The Equivalence of Elementary Particle Theories and Computer Languages: Quantum Computers, Turing Machines, Standard Model, Superstring Theory, and a Proof that Gödel's Theorem Implies Nature Must Be Quantum* (Pingree-Hill Publishing, Auburn, NH, 2005).

______, 2006a, *The Foundation of the Forces of Nature* (Pingree-Hill Publishing, Auburn, NH, 2006).

______, 2006b, *A Derivation of ElectroWeak Theory based on an Extension of Special Relativity; Black Hole Tachyons; & Tachyons of Any Spin.* (Pingree-Hill Publishing, Auburn, NH, 2006).

_______, 2007a, *Physics Beyond the Light Barrier: The Source of Parity Violation, Tachyons, and A Derivation of Standard Model Features* (Pingree-Hill Publishing, Auburn, NH, 2007).

______, 2007b, *The Origin of the Standard Model: The Genesis of Four Quark and Lepton Species, Parity Violation, the ElectroWeak Sector, Color SU(3), Three Visible Generations of Fermions, and One Generation of Dark Matter with Dark Energy* (Pingree-Hill Publishing, Auburn, NH, 2007).

______, *2008a, A Direct Derivation of the Form of the Standard Model From GL(16) (Pingree-Hill Publishing, Auburn, NH, 2008).*

______, 2008b, *A Complete Derivation of the Form of the Standard Model With a New Method to Generate Particle Masses Second Edition* (Pingree-Hill Publishing, Auburn, NH, 2008)

______, 2009, *The Algebra of Thought & Reality: The Mathematical Basis for Plato's Theory of Ideas, and Reality Extended to Include A Priori Observers and Space-Time Second Edition* (Pingree-Hill Publishing, Auburn, NH, 2009).

______, 2010a, *Operator Metaphysics: A New Metaphysics Based on a New Operator Logic and a New Quantum Operator Logic that Lead to a Mathematical Basis for Plato's Theory of Ideas and Reality* (Pingree-Hill Publishing, Auburn, NH, 2010).

______, 2010b, *The Standard Model's Form Derived from Operator Logic, Superluminal Transformations and GL(16)* (Pingree-Hill Publishing, Auburn, NH, 2010).

______, 2010c, *SuperCivilizations: Civilizations as Superorganisms* (McMann-Fisher Publishing, Auburn, NH, 2010).

______, 2011a, *21st Century Natural Philosophy Of Ultimate Physical Reality* (McMann-Fisher Publishing, Auburn, NH, 2011).

______, 2011b, *All the Universe! Faster Than Light Tachyon Quark Starships & Particle Accelerators with the LHC as a Prototype Starship Drive Scientific Edition* (Pingree-Hill Publishing, Auburn, NH, 2011).

______, 2011c, *From Asynchronous Logic to The Standard Model to Superflight to the Stars* (Blaha Research, Auburn, NH, 2011).

______, 2012a, *From Asynchronous Logic to The Standard Model to Superflight to the Stars volume 2: Superluminal CP and CPT, U(4) Complex General Relativity and The Standard Model, Complex Vierbein General Relativity, Kinetic Theory, Thermodynamics* (Blaha Research, Auburn, NH, 2012).

______, 2012b, *Standard Model Symmetries, And Four And Sixteen Dimension Complex Relativity; The Origin Of Higgs Mass Terms* (Blaha Reasearch, Auburn, NH, 2012).

______, 2013a, *Multi-Stage Space Guns, Micro-Pulse Nuclear Rockets, and Faster-Than-Light Quark-Gluon Ion Drive Starships* (Blaha Research, Auburn, NH, 2013).

______, 2013b, *The Bridge to Dark Matter; A New Sister Universe; Dark Energy; Inflatons; Quantum Big Bang; Superluminal Physics; An Extended Standard Model Based on Geometry* (Blaha Reasearch, Auburn, NH, 2013).

______, 2014a, *Universes and Megaverses: From a New Standard Model to a Physical Megaverse; The Big Bang; Our Sister Universe's Wormhole; Origin of the Cosmological Constant, Spatial Asymmetry of the Universe, and its Web of Galaxies; A Baryonic Field between Universes and Particles; Megaverse Extended Wheeler-DeWitt Equation* (Blaha Reasearch, Auburn, NH, 2014).

______, 2014b, *All the Megaverse! Starships Exploring the Endless Universes of the Cosmos Using the Baryonic Force* (Blaha Research, Auburn, NH, 2014).

______, 2014c, *All the Megaverse! II Between Megaverse Universes: Quantum Entanglement Explained by the Megaverse Coherent Baryonic Radiation Devices – PHASERs Neutron Star Megaverse Slingshot Dynamics Spiritual and UFO Events, and the Megaverse Microscopic Entry into the Megaverse* (Blaha Research, Auburn, NH, 2014).

______, 2015a, *PHYSICS IS LOGIC PAINTED ON THE VOID: Origin of Bare Masses and The Standard Model in Logic, U(4) Origin of the Generations, Normal and Dark Baryonic Forces, Dark Matter, Dark Energy, The Big Bang, Complex General Relativity, A Megaverse of Universe Particles* (Blaha Research, Auburn, NH, 2015).

______, 2015b, *PHYSICS IS LOGIC Part II: The Theory of Everything, The Megaverse Theory of Everything, U(4)⊗U(4) Grand Unified Theory (GUT), Inertial Mass = Gravitational Mass, Unified Extended Standard Model and a New Complex General Relativity with Higgs Particles, Generation Group Higgs Particles* (Blaha Research, Auburn, NH, 2015).

______, 2015c, *The Origin of Higgs ("God") Particles and the Higgs Mechanism: Physics is Logic III, Beyond Higgs – A Revamped Theory With a Local Arrow of Time, The Theory of Everything Enhanced, Why Inertial Frames are Special, Universes of the Mind* (Blaha Research, Auburn, NH, 2015).

______, 2015d, *The Origin of the Eight Coupling Constants of The Theory of Everything: U(8) Grand Unified Theory of Everything (GUTE), S^8 Coupling Constant Symmetry, Space-Time Dependent Coupling Constants, Big Bang Vacuum Coupling Constants, Physics is Logic IV* (Blaha Research, Auburn, NH, 2015).

______, 2016a, *New Types of Dark Matter, Big Bang Equipartition, and A New U(4) Symmetry in the Theory of Everything: Equipartition Principle for Fermions, Matter is 83.33% Dark, Penetrating the Veil of the Big Bang, Explicit QFT Quark Confinement and Charmonium, Physics is Logic V* (Blaha Research, Auburn, NH, 2016).

______, 2016b, *The Periodic Table of the 192 Quarks and Leptons in The Theory of Everything: The U(4) Layer Group, Physics is Logic VI* (Blaha Research, Auburn, NH, 2016).

______, 2016c, *New Boson Quantum Field Theory, Dark Matter Dynamics, Dark Matter Fermion Layer Mixing, Genesis of Higgs Particles, New Layer Higgs Masses, Higgs Coupling Constants, Non-Abelian Higgs Gauge Fields, Physics is Logic VII* (Blaha Research, Auburn, NH, 2016).

______, 2016d, *Unification of the Strong Interactions and Gravitation: Quark Confinement Linked to Modified Short-Distance Gravity; Physics is Logic VIII* (Blaha Research, Auburn, NH, 2016).

______, 2016e, *MoND: Unification of the Strong Interactions and Gravitation II, Quark Confinement Linked to Large-Scale Gravity, Physics is Logic IX* (Blaha Research, Auburn, NH, 2016).

______, 2016f, *CQMechanics: A Unification of Quantum & Classical Mechanics, Quantum/Semi-Classical Entanglement, Quantum/Classical Path Integrals, Quantum/Classical Chaos* (Blaha Research, Auburn, NH, 2016).

______, 2016g, *GEMS: Unified Gravity, ElectroMagnetic and Strong Interactions: Manifest Quark Confinement, A Solution for the Proton Spin Puzzle, Modified Gravity on the Galactic Scale* (Pingree Hill Publishing, Auburn, NH, 2016).

______, 2016h, *Unification of the Seven Boson Interactions based on the Riemann-Christoffel Curvature Tensor* (Pingree Hill Publishing, Auburn, NH, 2016).

______, 2017a, *Unification of the Eleven Boson Interactions based on 'Rotations of Interactions'* (Pingree Hill Publishing, Auburn, NH, 2017).

______, 2017b, *The Origin of Fermions and Bosons,and Their Unification* (Blaha Research, Auburn, NH, 2017).

Eddington, A. S., 1952, *The Mathematical Theory of Relativity* (Cambridge University Press, Cambridge, U.K., 1952).

Fant, Karl M., 2005, *Logically Determined Design: Clockless System Design With NULL Convention Logic* (John Wiley and Sons, Hoboken, NJ, 2005).

Feinberg, G. and Shapiro, R., 1980, *Life Beyond Earth: The Intelligent Earthlings Guide to Life in the Universe* (William Morrow and Company, New York, 1980).

Gelfand, I. M., Fomin, S. V., Silverman, R. A. (tr), 2000, *Calculus of Variations* (Dover Publications, Mineola, NY, 2000).

Giaquinta, M., Modica, G., Souchek, J., 1998, *Cartesian Coordinates in the Calculus of Variations* Volumes I and II (Springer-Verlag, New York, 1998).

Giaquinta, M., Hildebrandt, S., 1996, *Calculus of Variations* Volumes I and II (Springer-Verlag, New York, 1996).

Gradshteyn, I. S. and Ryzhik, I. M., 1965, *Table of Integrals, Series, and Products* (Academic Press, New York, 1965).

Heitler, W., 1954, *The Quantum Theory of Radiation* (Claendon Press, Oxford, UK, 1954).

Huang, Kerson, 1992, *Quarks, Leptons & Gauge Fields 2nd Edition* (World Scientific Publishing Company, Singapore, 1992).

Jost, J., Li-Jost, X., 1998, *Calculus of Variations* (Cambridge University Press, New York, 1998).

Landau, L. D. and Lifshitz, E. M., 1987, *Fluid Mechanics 2nd Edition*, (Pergamon Press, Elmsford, NY, 1987).

Misner, C. W., Thorne, K. S., and Wheeler, J. A., 1973, *Gravitation* (W. H. Freeman, New York, 1973).

Rescher, N., 1967, *The Philosophy of Leibniz* (Prentice-Hall, Englewood Cliffs, NJ, 1967).

Sagan, H., 1993, *Introduction to the Calculus of Variations* (Dover Publications, Mineola, NY, 1993).

Sakurai, J. J., 1964, *Invariance Principles and Elementary Particles* (Princeton University Press, Princeton, NJ, 1964).

Streater, R. F. and Wightman, A. S., 2000, *PCT, Spin, Statistics, and All That* (Princeton University Press, Princeton, NJ 2000).

Weinberg, S., 1972, *Gravitation and Cosmology* (John Wiley and Sons, New York, 1972).

Weinberg, S., 1995, *The Quantum Theory of Fields Volume I* (Cambridge University Press, New York, 1995).

Weyl, H., 1950, *Space, Time, Matter* (Dover, New York, 1950).

Weyl, H., (Tr. S. Pollard et al), 1987, *The Continuum* (Dover Publications, New York, 1987).

INDEX

About the Author

Stephen Blaha is a well known Physicist and Man of Letters with interests in Science, Society and civilization, the Arts, and Technology. He had an Alfred P. Sloan Foundation scholarship in college. He received his Ph.D. in Physics from Rockefeller University. He has served on the faculties of several major universities. He was also a Member of the Technical Staff at Bell Laboratories, a manager at the Boston Globe Newspaper, a Director at Wang Laboratories, and President of Blaha Software Inc and of Janus Associates Inc. (NH).

Among other achievements he was a co-discoverer of the “r potential” for heavy quark binding developing the first (and still the only demonstrable) non-abelian gauge theory with an “r” potential; first suggested the existence of topological structures in superfluid He-3; first proposed Yang-Mills theories would appear in condensed matter phenomena with non-scalar order parameters; first developed a grammar-based formalism for quantum computers and applied it to elementary particle theories; first developed a new form of quantum field theory without divergences (thus solving a major 60 year old problem that enabled a unified theory of the Standard Model and Quantum Gravity without divergences to be developed); first developed a formulation of complex General Relativity based on analytic continuation from real space-time; first developed a generalized non-homogeneous Robertson-Walker metric that enabled a quantum theory of the Big Bang to be developed without singularities at t = 0; first generalized Cauchy’s theorem and Gauss’ theorem to complex, curved multi-dimensional spaces; received Honorable Mention in the Gravity Research Foundation Essay Competition in 1978; first developed a physically acceptable theory of faster-than-light particles; first derived a composition of extrema method in the Calculus of Variations; first quantitatively suggested that inflationary periods in the history of the universe were not needed; first proved Gödel’s Theorem implies Nature must be quantum; provided a new alternative to the Higgs Mechanism, and Higgs particles, to generate masses; first showed how to resolve logical paradoxes including Gödel’s Undecidability Theorem by developing Operator Logic and Quantum Operator Logic; first developed a quantitative harmonic oscillator-like model of the life cycle, and interactions, of civilizations; first showed how equations describing superorganisms also apply to civilizations. A recent book shows his theory applies successfully to the past 14 years of history and to *new* archaeological data on Andean and Mayan civilizations as well as Early Anatolian and Egyptian civilizations.

He first developed an axiomatic derivation of the forms of The Standard Model from geometry – space-time properties – The Extended Standard Model. It has a Dark Matter sector that approximates the ElectroWeak sector with Dark doublets and Dark gauge interactions. It also uses quantum coordinates to remove infinities that crop up in most interacting quantum field theories and additionally to remove the infinities that appear in the Big Bang and generate an inflationary growth of the universe. The Extended Standard Model has an ultra-high energy

GUT (Grand Unified Theory) limit with a U(4)⊗U(4) symmetry; and can be united with gravitation to form a Theory of Everything. (See *Physics is Logic Part II.*)

Blaha has had a major impact on a succession of elementary particle theories: his Ph.D. thesis (1970), and papers, showed that quantum field theory calculations to all orders in ladder approximations could not give scaling deep inelastic electron-nucleon scattering. He later showed the eigenvalue equation for the fine structure constant α in Johnson-Baker-Willey QED had a zero at $\alpha = 1$ not 1/137 by solving the Schwinger-Dyson equations to all orders in an approximation that agreed with exact results to 4th order in α thus ending interest in this theory. In 1979 at Prof. Ken Johnson's (MIT) suggestion he calculated the proton-neutron mass difference in the MIT bag model and found the result had the wrong sign reducing interest in the bag model. These results all appear in Physical Review papers. In the 2000's he repeatedly pointed out the shortcomings of SuperString theory and showed that The Standard Model's form could be derived from space-time geometry by an extension of Lorentz transformations to faster than light transformations. This deeper space-time basis greatly increases the possibility that it is part of THE fundamental theory.Recently, Blaha showed that the Weak interactions differed significantly from the Strong, electromagnetic and gravitation interactions in important respects while these interactions had similar features, and suggested that ElectroWeak theory, which is essentially a glued union of the Weak interactions and Electromagnetism, possibly modulo unknown Higgs particle features, be replaced by a unified theory of the other interactions combined with a stand-alone Weak interaction theory. Blaha also showed that, if Charmonium calculations are taken seriously, the Strong interaction coupling constant is only a factor of five larger than the electromagnetic coupling constant, and thus Strong interaction perturbation theory would make sense and yield physically meaningful results.

In graduate school (1965-71) he wrote substantial papers in elementary particles and group theory: The Inelastic E- P Structure Functions in a Gluon Model. Phys. Lett. B40:501-502,1972; Deep-Inelastic E-P Structure Functions In A Ladder Model With Spin 1/2 Nucleons, Phys.Rev. D3:510-523,1971; Continuum Contributions To The Pion Radius, Phys. Rev. 178:2167-2169,1969; Character Analysis of U(N) and SU(N), J. Math. Phys. 10, 2156 (1969); and The Calculation of the Irreducible Characters of the Symmetric Group in Terms of the Compound Characters, (Published as Blaha's Lemma in D. E. Knuth's book: *The Art of Computer Programming Vols. 1 – 4*).

In the early 1980's Blaha was also a pioneer in the development of UNIX for financial, scientific and Internet applications: benchmarked UNIX versions showing that block size was critical for UNIX performance, developing financial modeling software, starting database benchmarking comparison studies, developing Internet-like UNIX networking (1982) and developing a hybrid shell programming technique (1982) that was a precursor to the PERL programming language. He was also the manager of the AT&T ten-year future products development database. His work helped lead to commercial UNIX on computers such as Sun Micros, IBM AIX minis, and Apple computers.

In the 1980's he pioneered the development of PC Desktop Publishing on laser printers. and was nominated for three "Awards for Technical Excellence" in 1987 by PC Magazine for PC software products that he designed and developed.

Recently he has developed a theory of Megaverses – actual universes of which our universe is one – with quantum particle-like properties based on the Wheeler-DeWitt equation of Quantum Gravity. He has developed a theory of a baryonic force, which had been conjectured many years ago, and estimated the strength of the force based on discrepancies in measurements of the gravitational constant G. This force, operative in 15-dimensinal space, can be used to escape from our universe in "uniships" which are the equivalent of the faster-than-light starships proposed in the author's earlier books. Thus travel to other universes, as well as to other stars is possible.

Blaha also considered the complexified Wheeler-DeWitt equation and showed that its limitation to real-valued coordinates and metrics generated a Cosmological Constant in the Einstein equations.

The author has also recently written a series of books on the serious problems of the United States and their solution as well as a book on the decline of Mankind that will follow from current social and genetic trends in Mankind.

In the past twelve years Dr. Blaha has written over 40 books on a wide range of topics. Some recent major works are: *From Asynchronous Logic to The Standard Model to Superflight to the Stars*, *All the Universe!*, *SuperCivilizations: Civilizations as Superorganisms, America's Future: an Islamic Surge, ISIS, al Qaeda, World Epidemics, Ukraine, Russia-China Pact, US Leadership Crisis,The Rises and Falls of Man – Destiny – 3000 AD: New Support for a Superorganism MACRO-THEORY of CIVILIZATIONS From CURRENT WORLD TRENDS and NEW Peruvian, Pre-Mayan, Mayan, Anatolian, and Early Egyptian Data, with a Projection to 3000 AD,* and *Mankind in Decline: Genetic Disasters, Human-Animal Hybrids, Overpopulation, Pollution, Global Warming, Food and Water Shortages, Desertification, Poverty, Rising Violence, Genocide, Epidemics, Wars, Leadership Failure.*

He has taught approximately 4,000 students in undergraduate, graduate, and postgraduate corporate education courses primarily in major universities, and large companies and government agencies.

The above paragraphs summarize much of his work over the past fifty years. This work is fully documented. He continues to engage in research and writing at Blaha Research.

www.ingramcontent.com/pod-product-compliance
Lightning Source LLC
Chambersburg PA
CBHW082003190326
41458CB00010B/3056